Memory

Memory
From Mind to Molecules

Larry R. Squire
Eric R. Kandel

**SCIENTIFIC
AMERICAN
LIBRARY**

A division of HPHLP
New York

Jacket and Text Designer: *Cambraia Magalhães*

Library of Congress Cataloging-in-Publication Data

Squire, Larry R.
  Memory : from mind to molecules / Larry R. Squire, Eric R. Kandel.
    p.   cm.
  Includes bibliographical references and index.
  ISBN 0-7167-5071-6 (hardcover)   ISBN 0-7167-6037-1 (paperback)
  1. Memory. I. Kandel, Eric R.   II. Title.
  QP406.S663   1999
612.8′2 — dc21   98-44075
                                                             CIP

ISSN 1040-3213

Printed in the United States of America

Scientific American Library
A division of HPHLP
New York

Distributed by W. H. Freeman and Company
41 Madison Avenue, New York, NY 10010
Houndmills, Basingstoke RG21 6XS, England

First printing, November 1998

To
the Squire children: Ryan and Luke
and
the Kandel grandchildren: Allison, Libby, and Izak

Contents

Preface

"Cogito ergo sum." — "I think therefore I am." This phrase, written by the great French philosopher René Descartes in 1637, still stands as perhaps the most widely quoted pronouncement in all of Western philosophy. One of the great biological lessons of the twentieth century, and a lesson that serves as the starting point of this book, is that the statement is wrong and that it is wrong for two reasons. First, Descartes used this phrase to emphasize the separation he believed to exist between the mind and the body. He considered mental activity to be completely independent of the activity of the body. However, biologists now have every reason to believe that all the activities of the mind arise from a specialized part of our body: our brain. Accordingly, it would be more correct to rephrase Descartes' statement by reversing it to read "I am, therefore I think" — much as the neurologist Antonio Damasio suggests in his charming book, *Descartes' Error*. In modern terms we would say "I have a brain, therefore I think."

There is, however, a second and larger sense in which Descartes' original statement is wrong. We are not who we are simply because we think. We are who we are because we can remember what we have thought about. As we try to illustrate in the chapters that follow, every thought we have, every word we speak, every action we engage in — indeed, our very sense of self and our sense of connectedness to others — we owe to our memory, to the ability of our brains to record and store our experiences. Memory is the glue that binds our mental life, the scaffolding that holds our personal history and that makes it possible to grow and change throughout life. When memory is lost, as in Alzheimer's disease, we lose the ability to re-create our past, and as a result we lose our connection with ourselves and with others.

During the last two decades, there has been a revolution in our understanding of memory and what happens in the brain when we learn and remember. Our purpose in writing this book is to outline the exciting origins of this revolution and to describe what we now know about the workings of memory, and the workings of nerve cells and brain systems. We also describe how memory can go awry after injury or disease.

The modern study of memory has two sources. The first is the biological study of how nerve cells signal one another. The key discovery here is that signaling by nerve cells is not fixed but can be modulated by activity and by experience. Thus experience can leave a record in the brain, and it does so by using nerve cells as elementary memory storage devices. The second source is the study of brain systems and cognition. The key discovery here is that memory is not unitary but has different forms that use distinctive logic and different brain circuits. In this book we have tried to combine these two historically disparate strands to create a new synthesis: *a molecular biology of cognition*, which emphasizes the interplay between the molecular biology of signaling and the cognitive neuroscience of memory.

Some of the advances that we here describe have come from studies of memory within the neuronal circuits of simple invertebrate animals; other advances have come from studies of more complex nervous systems, including the human brain. Among the new technical developments that have moved this work forward is the ability to image the human brain while people learn and remember, and the availability of genetic techniques for studying memory in intact animals such as mice.

The book reflects our view that the time is ripe for a treatment of memory that covers the full range of the topic—from mind to molecule. In writing this book, we have been guided by our own perspectives on memory and have not tried to provide an exhaustive coverage of the field. Rather, we have emphasized research that we have been involved in directly or that has directly influenced our thinking. Each chapter of the book represents our joint synthesis of the current state of knowledge. The first and last chapters were written jointly, and each of the other chapters was also a collaborative effort. Eric Kandel initially drafted the chapters that focus on cell and molecular storage mechanisms (2, 3, 6, and 7), and Larry Squire drafted the chapters that focus on cognition and brain systems (4, 5, 8, and 9). Every chapter was then subjected to extensive comment and revision by the other coauthor. These extended, critical exchanges led to a final version that is better than either of us could have achieved alone. In a larger sense, the book is a reflection of the continuing dialogue we have enjoyed over the past three decades and the friendship that has grown out of it.

We wrote this book for a broad audience. Foremost in our minds was the general reader who enjoys science and is interested in becoming acquainted with the remarkable new discoveries about how the nervous system learns and remembers. Because we were specifically interested in reaching nonscientific readers who do not have specialized backgrounds for the topics of this book, we have provided where necessary a rudimentary background for the relevant biology and cognitive psychology. In addition, precisely because we wrote with the nonspecialist in mind, the book should be useful to undergraduate and graduate students. We believe this to be the first treatment of memory that extends from cognition to molecular biology, and we would hope that students will find here a clear and readable introduction to the topic. Finally, our scientific colleagues in psychology and neuroscience, as well as instructors in undergraduate and graduate courses, may find the book informative and useful as a single-volume overview of what is currently an extremely active and exciting area of scholarship and research.

We have been greatly aided in our task by the staff at W. H. Freeman and Company. We are especially grateful to Jonathan Cobb, who facilitated our collaboration and who read and revised every chapter with sensitive editorial insights, even when he no longer had responsibility for doing so. We have also benefited enormously from the thoughtful editing and constructive suggestions of our other editor, Susan Moran, and from the advice of Jane O'Neill, our project editor. Finally, we owe a special debt to the colleagues who graciously read and commented on various sections of the book: Ted Abel, Marilyn Albert, David Amaral, Craig Bailey, Jack Byrne, Michael Davis, Joaquin Fuster, Emma Gibbs, Jon Kaas, Sarah Mack, Alex Martin, Chris Pittenger, Richard Thompson, Leslie Ungerleider, Torsten Wiesel, and Stuart Zola.

Memory

Marc Chagall, The Birthday *(1915). The subject matter of the paintings of Chagall (1887–1985) is often romantic and dreamlike, with references that recall his traditional Jewish upbringing in Russia. Here he captures the joy and memory of his early love.*

1 | From Mind to Molecules

Memory collects the countless phenomena of our existence into a single whole . . . our consciousness would be broken up into as many fragments as we had lived seconds but for the binding and unifying force of memory.

— Ewald Hering

E.P. had been a successful laboratory technician for 28 years when he retired in 1982 to enjoy his family and pursue his hobbies. Ten years later, at the age of 70, E.P. suddenly developed an acute viral illness—herpes simplex encephalitis—that required hospitalization. When he returned home from the hospital, his friends and family saw the same high-spirited and friendly man they had always known. He smiled easily and liked to laugh and tell stories. He looked in fine physical health, his walk and manner were as before, and his voice was strong and clear. He was alert and attentive, and he conversed appropriately with visitors. Indeed, testing would later show that his thought processes were intact. But it took only a few moments to sense that something was very wrong with his memory. He repeated the same comments and asked the same questions over and over again, and he could not keep up with conversations. He never came to recognize new visitors to his house, even after 40 visits.

The herpes simplex virus had destroyed parts of E.P.'s brain, and with this brain damage he had lost his ability to make new memories. He could now remember new events or encounters for only a few seconds. Looking backward from the present, he was uncertain which house he had earlier lived in for 20 years, or that one of his grown children lived next door, or that he had two grandchildren. The illness had kept him from carrying his thoughts and impressions into the future, and it had broken his connection to the past, to what had happened in his life before. He was now confined, in a manner of speaking, to the present, to the immediate moment.

As the consequences of E.P.'s viral encephalitis make dramatically clear, learning and memory are fundamental to human experience. We can acquire new knowledge about the world because the experiences we have modify our brain. And, once learned, we can hold the new knowledge in our memories, often for a very long time, because aspects of the modifications are retained in our brains. At a later time, we can act on the knowledge that is stored in memory, behaving and thinking in new ways. Memory is the process by which what is learned persists across time. In this sense, learning and memory are inextricably connected.

Most of what we know about the world is not built into our brains at birth but is acquired through experience and maintained through memory—the names and faces of our friends and loved ones, algebra and geography, politics and sports, and the music of Haydn, Mozart, and Beethoven. As a result, we are who we are in large part because of what we learn and remember. But memory is not only a record of personal experience: it also allows us to become educated, and it is a powerful force for social progress. Humans have the unique ability to communicate what they have learned to others, and in so doing they can create cultures that can

be transmitted from generation to generation. Human accomplishment seems ever expanding, yet the size of the human brain does not appear to have increased significantly since *Homo sapiens* first appeared in the fossil record several hundred thousand years ago. What has determined cultural change and progress during these many thousands of years is not an increase in brain size or even a change in brain structure. It is rather the intrinsic capacity of the human brain to capture what we learn in speech and writing, and to teach it to others.

While memory is central to many of the most positive aspects of human experience, it is also true that many psychological and emotional problems result, at least in part, from experience that is coded in memory. These problems are learned, often in response to experiences of early life that result in habitual ways of interacting with the world. Moreover, insofar as psychotherapeutic intervention is successful in treating mental disorders, it presumably succeeds by creating experiences that allow people to learn anew.

Loss of memory leads to the loss of self, the loss of one's life history, and the loss of enduring interactions with other human beings. Disorders of learning and disturbances of memory haunt the developing infant as well as the mature adult. Mental retardation, Down's syndrome, dyslexia, the normal weakening of memory with age, and the devastation of Alzheimer's disease and Huntington's disease are only the more familiar examples of a large number of diseases that affect memory.

The analysis of how learning occurs and how memories are stored has been central to three intellectual disciplines: first philosophy, then psychology, and now biology. Until late in the nineteenth century, the study of memory was restricted largely to the domain of philosophy. However, during the twentieth century, the focus of inquiry gradually moved to more experimental studies, initially in psychology and more recently in biology. As we enter the next millennium, the questions posed by psychology and biology have begun to converge on common ground. From the perspective of psychology these questions are: How does memory work? Are there different kinds of memory? If so, what is their logic? From the perspective of biology these questions are: Where in the brain do we learn? Where do we store what is learned as memory? Can memory storage be resolved at the level of individual nerve cells? If so, what is the nature of the molecules that underlie the various processes of memory storage? Neither psychology nor biology alone can satisfactorily address these questions, but the combined strength of both disciplines is providing a fresh and exciting picture of how the brain learns and remembers. Psychologists and biologists have together defined a common program of inquiry, centered around two sorts of overriding questions: (1) How are the various forms of memory organized in the brain? (2) How is memory storage accomplished? Our purpose in this book is to address these questions.

The convergence of psychology and biology has led to a new synthesis of knowledge about learning and memory. We now know that there are many forms of memory, that different brain structures carry out specific jobs, and that memory is encoded in individual nerve cells and depends on changes in the strength of their interconnections. We also know that these changes are stabilized by the actions of genes in nerve cells, and we know something about how the molecules inside nerve cells change the connection strength between nerve cells. Memory promises to be the first mental faculty to be understandable in a language that makes a bridge from molecules to mind, that is, from molecules to cells, to brain systems, and to behavior. This developing understanding, in turn, is likely to lead to new insights into the causes and treatment of memory disorder.

MEMORY AS A PSYCHOLOGICAL PROCESS

Since Socrates first proposed that humans have preknowledge—that certain knowledge about the world is inborn—Western philosophy has struggled with several related questions: How do we learn new information about the world, and how does this information become stored in memory? What aspects of the mind's knowledge are innate, and to what extent can experience influence that innate organization? Initially, philosophers used essentially three methods—all non-experimental—for studying memory and other mental processes: conscious introspection, logical analysis, and argument. The difficulty was that these methods did not lead to agreed upon facts or build consensus toward a common point of view. By the middle of the nineteenth century, the success of experimental science in solving problems in physics and chemistry began to attract students of behavior and the mind. As a result, the philosophical exploration of mental processes was gradually replaced by empirical studies of the mind, and psychology emerged as an independent discipline, distinct from philosophy.

At the beginning, experimental psychologists focused their studies on sense perception, but gradually they ventured into the more complex workings of the mind and tried to submit mental phenomena to experimental and quantitative analysis. The pioneer in this effort was the German psychologist Hermann Ebbinghaus, who in the 1880s succeeded in bringing the study of memory into the laboratory. To study memory objectively and quantitatively, Ebbinghaus wanted to use standardized, homogeneous test

Hermann Ebbinghaus (1850–1909), German psychologist who introduced experimental methods into psychology and pioneered the laboratory study of learning and memory.

items that he could ask a subject to memorize. For this purpose he invented a new type of syllable in which a vowel sound is placed between two consonants, as in BIK or REN. He constructed about 2300 of these, wrote them on individual pieces of paper, mixed them together, and then drew them out at random to form lists for his learning experiments. Using himself as a subject, he learned lists of these syllables and then tested his recollection at various time intervals afterward. He also measured the number of repetitions required and the time needed to relearn each list.

By this means Ebbinghaus was able to discover two key principles of memory storage. First, he showed that memories have different life spans. Some memories are short-lived and retained for minutes; others are long-lived and persist for days to months. Second, he showed that repetition makes memories last longer—it is practice that makes perfect. With a single practice session, a list might be remembered for only a few minutes, but with enough repetition the list could persist in memory for days and weeks. A few years later, the German psychologists Georg Müller and Alfons Pilzecker suggested that this memory that lasts days and weeks becomes *consolidated* with time. A memory that has become consolidated is robust and resistant to interference. In its initial stages, even memory that would otherwise persist is highly susceptible to disruption if, for example, an attempt is made to learn some other similar material.

The American philosopher William James later elaborated these findings by making a sharp, qualitative distinction between short-term and long-term memory. Short-term memory, he argued, lasts seconds to minutes and is essentially an extension of the present moment, as when one looks up a telephone number and then holds it in mind for a moment. By contrast, long-term memory can last weeks, months, even a lifetime, and is consulted by reaching back into the past. This distinction has proven fundamental to understanding memory.

At about the same time that Ebbinghaus and James did their classic work, the Russian psychiatrist Sergei Korsakoff published the first descriptions of a memory disorder that would eventually bear his name, Korsakoff's syndrome. It is now the best-known and most widely studied example of human amnesia. Even before Korsakoff's time, it was appreciated that the study of impaired memory could provide superb insight into the structure and organization of normal memory. As in other areas of biology, where the analysis of disease has helped to illuminate normal function, so in memory it became clear that detailed study of memory impairments could provide a great deal of useful information. For example, the study of amnesia showed that there are multiple kinds of memory, a theme that will be emphasized often in this book.

THE BEHAVIORIST REVOLUTION

In the mid-nineteenth century, Charles Darwin suggested that mental characteristics have continuity across species in the same way as morphological features. Limbs, for example, are constructed on the same general pattern across mammals, birds, and reptiles, such that the forelimb of a lizard, the wing of a bat, and a human arm contain the same bones and relative arrangement of parts. If humans are similar to other animals in such important ways, we should be able to learn about our mental lives by studying animals. In the early twentieth century, following the success of Ebbinghaus's study of human memory and inspired by Darwin's idea that human mental capabilities have evolved from those of simpler animals, the noted Russian physiologist Ivan Pavlov and the American psychologist Edward Thorndike developed animal models for studying learning. Working independently, each discovered a different experimental method for modifying behavior: Pavlov discovered classical conditioning and Thorndike discovered instrumental, or operant, conditioning (known more familiarly as trial-and-error learning). These two experimental methods provided the foundation for the scientific study of animal learning and memory. In classical conditioning, an animal learns to associate two events, perhaps the sound of a bell and the presentation of food, so that the animal comes to salivate whenever the bell sounds, even in the absence of food. The animal has learned that the bell predicts the coming of food. In instrumental conditioning, an animal learns to make an association between a correct response and a reward, or an incorrect response and a punishment that follows the response, and in this way gradually modifies its behavior.

This objective, laboratory-based learning psychology developed into an empirical tradition called behaviorism, which changed how the study of memory was conducted. Behaviorists, led by the American John B. Watson, argued that behavior could now be studied with the same rigor as other natural sciences. Psychologists had only to focus exclusively on what was observable. They could identify stimuli and measure behavioral responses, but in this view the nature of an individual's experience and the nature of mental events could not be explored scientifically. Within this tradition the study of classical and instrumental conditioning yielded much useful information: lawful principles of how animals form associations between stimuli, the idea of reinforcement (or reward) as a key to understanding learning, and the description of how different schedules of reinforcement determine the rate of learning.

Despite its scientific rigor, behaviorism proved restrictive in its scope and limited in its method. In their attempt to emulate the natural sciences, and to study only observable stimuli and responses, the behaviorists lost sight of many other interesting and important questions about mental processes. In particular, they largely ignored the evidence from Gestalt psychology, neurology, psychoanalysis, and even everyday common sense, all of which pointed to the important mental machinery that intervenes between a stimulus and a response. Behaviorists essentially defined all of mental life in terms of the limited techniques that they used to study it. They restricted the domain of experimental psychology to a limited set of problems and excluded from study some of the most fascinating features of mental life, such as the cognitive processes that occur when we learn and remember. These intervening mental processes within the brain underlie perception, attention, motivation, action, planning, and thinking, as well as learning and memory.

THE COGNITIVE REVOLUTION

Behaviorism was the dominant psychological tradition in the study of learning and memory early in the twentieth century, especially in the United States. Yet there were a few notable departures from its orthodoxy, a few researchers for whom mental processes occupied center stage. One important forerunner to a less behavioral, and more cognitive, approach to memory was the British psychologist Frederic C. Bartlett. In the first half of the twentieth century, Bartlett studied memory in naturalistic settings by having people learn everyday material like stories and pictures. With these natural methods he demonstrated that memory is surprisingly fragile and susceptible to distortion. He suggested that memory retrieval is seldom highly exact. Retrieval is not simply a literal replaying of passively stored information waiting to be re-excited. Rather retrieval is essentially a creative, reconstructive process. His own words tell us:

> Remembering is not the re-excitation of innumerable fixed, lifeless and fragmentary traces. It is an imaginative reconstruction, or construction, built out of the relation of our attitude towards a whole active mass of organized past reactions or experience, and to a little outstanding detail which commonly appears in image or in language form.

By the 1960s, in no small part due to the work of Bartlett, the narrowness of behaviorism had become apparent to many psychologists. They came to the view that perception and memory depend not only on information in the environment but also on the mental structure of the perceiver or the rememberer. These ideas gave birth to the field of cognitive psychology. The important scientific task was to analyze not only stimuli and the responses they elicit but the processes that intervene between a stimulus and a behavior—precisely the domain ignored by behaviorists.

Frederic C. Bartlett (1886–1969), British psychologist and one of the founders of cognitive psychology. Bartlett added a naturalistic dimension to Ebbinghaus's rigorously controlled methods for studying memory.

In turning to the study of mental operations, cognitive psychologists attempted to follow the flow of information from the eye and the ear, and from the other sensory organs, to its *internal representation* in the brain for eventual use in memory and action. This internal representation was thought to take the form of a characteristic pattern of activity in particular groups of interconnected cells in the brain. Thus, when we view a scene, there is, the cognitive psychologists argued, a patterned activity in the brain that represents that scene.

Yet this new emphasis on internal representation was not without its own problems. Despite its narrowness, behaviorism was correct in emphasizing that internal representations are not readily accessible to objective analysis. Indeed, psychologists interested in cognition had to come to grips with the stern reality that internal representations of mental processes were tenuous theoretical constructs, difficult to approach experimentally. Reaction time measurements, for example, yielded insights about the order in which these hypothetical mental operations are carried out. However, these techniques examined mental operations indirectly and therefore could not indicate how an operation should be identified or what exactly it is. For cognitive psychology to flourish, it had to join forces with biology to open up the black box and explore the brain—so long ignored by the behaviorists.

THE BIOLOGICAL REVOLUTION

Fortunately, just as cognitive psychology was emerging in the 1960s a revolution was also occurring in biology that was bringing the interests of biology into a closer relationship with cognitive psychology. This revolution has had two major components: a molecular component and a systems component. Both of these have come to play a major role in the understanding of memory.

The molecular component of the biological revolution had its origin at the end of the nineteenth and the beginning of the twentieth century with the work of Gregor Mendel, William Bateson, and Thomas Hunt Morgan. These three showed that hereditary information is passed from parent to offspring by means of discrete biological units, which we now call genes, and that each gene resides at a specific locus on stringlike structures within the nucleus of the cell called *chromosomes*. In 1953, James Watson and Francis Crick solved the structure of DNA, the double-stranded molecule that constitutes the chromosomes and that contains the genes of all living organisms. This discovery led Crick to formulate the "central dogma" of molecular biology: that the DNA of genes contains a code (the genetic code) that can be *transcribed* into an intermediate molecule, messenger RNA, which in turn can be *translated* into protein.

In the late 1970s it became possible to read sequences of the genetic code readily and to see what protein a gene produces. It turned out that certain stretches of identical DNA encoded char-

acteristically recognizable domains or regions of proteins. Although these domains proved to be shared by a variety of different proteins, they mediated the same biological function. Thus, by looking at the coding *sequence* of a gene it became possible to infer aspects of the *function* of the protein it encodes. By merely comparing their sequences, one now could recognize relationships between proteins encountered in very different contexts: in different cells of the body of a given organism and even in vastly different organisms. As a result, there rapidly emerged a general blueprint for how cells function—and specifically how cells signal each other—that has provided a common conceptual framework for understanding many of life's processes. This framework has already had a major impact on the molecular study of learning in simple invertebrate animals such as the marine snail *Aplysia* and the fruit fly *Drosophila*. In the fullness of time, this framework should enable scientists to study, on a molecular level, the internal representation of complex cognitive processes in the mammalian brain.

The second, or systems, component of the biological revolution has been concerned with mapping elements of cognitive function onto specific brain areas. This component has been driven by the development of powerful methods for studying the internal representation of cognitive processes. Specifically, scientists now have the ability to record the activity of nerve cells in the brains of awake, behaving animals and to use PET scans and functional magnetic resonance to image the living human brain while a person is engaged in cognitive activity. Together these developments have made it possible to study what takes place in the brain when people perceive sensory stimuli, initiate motor action, and learn and remember.

As these several developments illustrate, the biology of memory can now be studied at two different levels, one aimed at nerve cells and the molecules within nerve cells, and the other aimed at brain structures, circuitry, and behavior. The

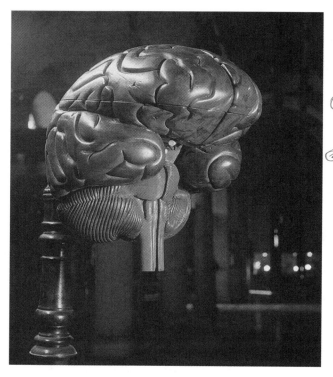

A wooden model of the human brain from the early nineteenth century.

the more general issue: Can any mental process be localized to a specific region or combination of regions in the brain? Since the beginning of the nineteenth century, two contrasting ideas have been advanced about the localization of mental function. One view is that the brain is composed of identifiable, localized parts and that language, vision, and other functions can be localized to particular regions. The other view is that the different mental functions are not localized to specific regions but instead are global properties that arise from the integrated activity of the entire brain. In a sense, the history of brain science can be seen as the gradual ascendancy of the first view, that the brain is composed of many different parts and that these parts are specialized for different functions—language, vision, and motor movements, for example.

first is concerned with the *cellular and molecular* mechanisms of memory storage. The second is concerned with the *neural systems* of the brain important for memory. Both approaches are yielding important insights about memory, and their synthesis promises to provide us with a new level of understanding. We first consider, in the next several sections, what has been learned at the level of neural systems.

THE NEURAL SYSTEMS FOR MEMORY: WHERE ARE MEMORIES STORED?

The question of where memories are stored is part of a long tradition that attempts to address

Karl Lashley (1890–1958), American psychologist who investigated the locus of memory in rats by removing different regions of the cerebral cortex.

Karl Lashly, mass action Theory, more cortex intact more memory, learning is not possible.

The person most identified with early attempts to localize memory in the brain was Karl Lashley, professor of psychology at Harvard University. In a series of famous experiments in the 1920s, Lashley trained rats to run through a simple maze. He then removed different areas of cortex, which is the most recently evolved, outer covering of the brain. He retested the rats 20 days later to see how much of their training they had retained. Based on these experiments, Lashley formulated the law of *mass action,* according to which the severity of memory impairment for the maze habit correlated with the size of the cortical area removed, not with its specific location. Thus, Lashley wrote:

> It is certain that the maze habit, when formed, is not localized in any single area of the cerebrum and that its performance is somehow conditioned by the quantity of tissue which is intact.

In 1950, toward the end of his career, Lashley summarized his search for the site of memory storage:

> This series of experiments has yielded a good bit of information about what and where the memory trace is not. It has discovered nothing directly of the real nature of the memory trace. I sometimes feel, in reviewing the evidence on the localization of the memory trace, that the necessary conclusion is that learning is just not possible. It is difficult to conceive of a mechanism that can satisfy the conditions set for it. Nevertheless, in spite of such evidence against it, learning sometimes does occur.

Study Problems with Lashlies

Many years later, after more experimental work, it was possible to arrive at a different understanding of Lashley's famous result. First, it became clear that Lashley's maze-learning task was unsuitable for studying localization of memory function because the task depended on many different sensory and motor capabilities. When a cortical lesion deprives an animal of one kind of

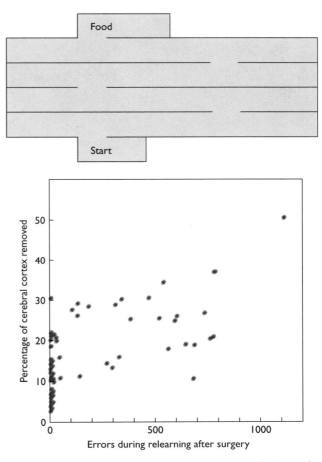

(Top) Ground plan of the rat maze used extensively by Lashley in his efforts to localize memory in the brain. (Bottom) Lashley found that the greater the extent of cortical damage in a rat, the more errors it made in relearning the maze.

cue (such as tactile cues), the animal can still remember reasonably well using its sense of vision or smell. In addition, Lashley focused his efforts only on the cerebral cortex, the outer layer of the brain. He did not explore structures that lie deeper in the brain, below the cortex. Subsequent work showed that many forms of memory require one or another of these *subcortical regions.* Nevertheless, Lashley's findings did rule out certain simple possibilities. For example, he showed that there is no single center in the brain where all memories are permanently stored. Many parts

Large areas of brain are involved in storage of memory but they are not equally involved.

Donald Hebb (1904–1985), Canadian psychologist. A colleague of Wilder Penfield and Brenda Milner, Hebb advocated the usefulness of understanding behavior in terms of brain function and emphasized the importance of distributed networks for memory storage.

of the brain must participate in the representation of memory.

An early response to Lashley's challenge about the locus of memory came from Donald O. Hebb, a psychologist at McGill University. To explain Lashley's result, that the connections formed by learning could not apparently be localized to a single brain region, Hebb suggested that assemblies of cells, distributed over large areas of cortex, work together to represent information. Within these assemblies enough interconnected cells will survive almost any lesion to ensure that information can still be represented.

Hebb's idea of a distributed memory store was farsighted. With the accumulation of addi-

tional evidence, this insight came to be seen as one of the key principles of information storage in the brain. No single memory region exists, and many parts of the brain participate in the representation of any single event. We now realize, however, that the idea of widely distributed memory storage is not the same as the idea that all the participating brain regions are equally involved in storing memory. The modern view is that memory is widely distributed but that different areas store different aspects of the whole. There is little redundancy or reduplication of function across these areas. Specific brain regions have specialized functions, and, as we will see in Chapter 5, each contributes in a different way to the storage of whole memories.

FIRST CLUES TO A LOCUS FOR MEMORY

The cerebral cortex is divided into four major regions or lobes. The frontal lobe is concerned with planning and with voluntary movement, the parietal lobe with sensations of the body surface and with spatial perception, the occipital lobe with vision, and the temporal lobe with hearing, visual perception, and, as we shall see, with memory.

The first suggestion that aspects of memory might be stored in the temporal lobe of the human brain came in 1938 from the work of an innovative neurosurgeon, Wilder Penfield. Working at the Montreal Neurological Institute, Penfield pioneered the neurosurgical treatment of focal epilepsy. This form of epilepsy produces brain seizures that are restricted to limited regions of cortex. Penfield developed a technique, still used, to remove the epileptic tissue while minimizing damage to the patient's mental functions. During surgery, he applied weak electrical stimulation to various sites in the cortex of his patients and determined its effects on the ability to speak and comprehend language. Because the brain contains no pain receptors, patients received only a local anesthetic; they remained fully conscious during surgery and were able to report their experiences. Through these responses, Penfield could identify specific brain sites important for language in the individual patient and then try to avoid these sites when removing epileptic tissue.

In this way, Penfield explored much of the cortical surface in more than 1000 patients. On occasion, he found that in response to electrical stimulation the patients would describe coherent perceptions or experiences. For example, one patient stated, "It sounded like a voice saying words, but it was so faint I couldn't get it." Another patient said, "I am seeing a picture of a dog and cat . . . the dog is chasing the cat." These responses were elicited invariably only from the temporal lobes of the brain, never from other areas, and even within the temporal lobes, stimulation evoked coherent experiences only rarely, in about 8 percent of cases. Nevertheless, these studies were fascinating in suggesting that the ex-

periences elicited by brain stimulation reproduced the stream of consciousness from a previous episode in the patient's life. However, this view eventually came into serious question. First, all the patients had abnormal brains due to their epilepsy, and in 40 percent of the cases the mental experience evoked by stimulation was identical to the mental experience that ordinarily accompanied the patient's seizures. Moreover, the mental experiences included elements of fantasy, as well as improbable or impossible situations, and were more like dreams than memories. In addition, removing the brain tissue under the stimulating electrode did not erase the memory of the elicited experience.

THE STORY OF AMNESIC PATIENT H.M.

Stimulated by Penfield's work, another neurosurgeon, William Scoville, soon obtained direct evidence that the temporal lobes are critically important for human memory. In 1957, Scoville and Brenda Milner, a psychologist at McGill University and a colleague of Penfield, reported the extraordinary story of patient H.M. At the age of about nine, H.M. was knocked down by a bicycle and sustained a head injury that led eventually to epilepsy. H.M.'s seizures worsened over the years until he was having as many as 10 blackouts and one major seizure every week. By the age of 27, he was severely incapacitated. Because H.M.'s epilepsy was thought to have its origin within the brain's temporal lobe, Scoville decided, as a last resort, to remove the inner surface of the temporal lobe on both sides of the brain, including a structure called the hippocampus, in an attempt to treat his epilepsy. This experimental treatment did help his epilepsy, but it left H.M. with a devastating memory loss from which he has never recovered. From the time of his operation in 1953 until the present day, H.M. has been

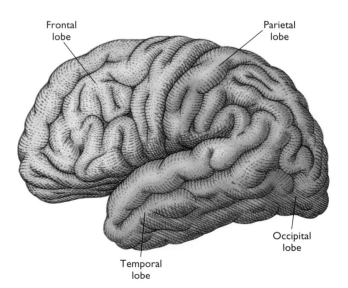

This lateral view of the human brain shows the four lobes of the left cerebral hemisphere.

Frontal lobe

Parietal lobe

Occipital lobe

Temporal lobe

lack of Hipocampus prevents conversion of STM to LTM.

Brenda Milner, the Canadian psychologist who studied H.M. and discovered the role of the medial temporal lobe in human memory.

unable to convert a new short-term memory into a permanent long-term memory.

Brenda Milner discovered this memory deficit and described it in a publication that became the most cited paper in the field of brain and behavior research. She has continued to follow H.M.'s case over the past 40 years. From the beginning, the most dramatic aspect of his impairment was that he appeared to forget events as soon as they happened. Whenever Brenda Milner entered the room to greet him, he failed to recognize her. Even now, less than an hour after eating, he cannot remember anything that he has eaten or even the fact that he has had a meal. As the years have passed, he can no longer recognize himself in a photograph because he has no memory of his changed appearance. Yet he is able to retain new information as long as his attention is not diverted from it.

H.M.'s memory deficits are remarkable. Consider his reaction to the task of remembering the number "584." Left to himself, he was able to re-

tain this information for several minutes by working out mnemonic schemes and holding the information continuously in mind. H.M. explained how he did it:

> It's easy. You just remember 8. You see 5, 8, and 4 add to 17. You remember 8, subtract it from 17 and it leaves 9. Divide 9 in half and you get 5 and 4 and there you are: 584. Easy.

Yet, only a minute or two later, after his attention had been directed to another task, he could not remember the number or any of his train of thought about the number. When H.M. was asked to try to remember the pair of words "nail" and "salad" by forming a visual image that combined the two words, he described in some detail how he had made a mental picture of a nail sticking into a salad and that he had tried to decide whether the head of the nail was up or down. He also said that he was making sure that the nail was large enough that he could see it and would not eat it by mistake. A few minutes later, he had forgotten the nail, the salad, and the imagery that he had constructed.

From these studies of H.M., Brenda Milner extracted four important principles. First, the ability to acquire new memories is a distinct cerebral function, localized to the medial portion of the temporal lobes of the brain, and separable from other perceptual and cognitive abilities. Thus, the brain has to some extent separated its perceptual and intellectual functions from its capacity to lay down in memory the records that ordinarily result from engaging in perceptual and intellectual work.

Second, the medial temporal lobes are not required for immediate memory. H.M. has perfectly good immediate memory. He can retain a number or a visual image for a short period after learning. He can also carry on a normal conversation, providing it does not last too long or move among too many topics.

Third, the medial temporal lobe and the hippocampus cannot be the ultimate storage sites for

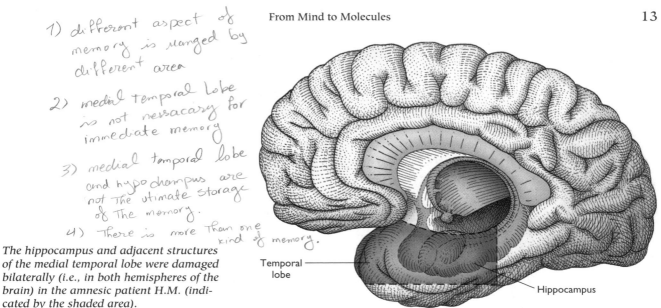

1) different aspect of memory is manged by different area

2) medial temporal Lobe is not nessacary for immediate memory

3) medial temporal lobe and hypochampus are not The utimate storage of The memory.

4) There is more Than one kind of memory.

Temporal lobe

Hippocampus

The hippocampus and adjacent structures of the medial temporal lobe were damaged bilaterally (i.e., in both hemispheres of the brain) in the amnesic patient H.M. (indicated by the shaded area).

long-term memory of previously acquired knowledge. H.M. can remember vividly many events from his childhood. (We now have reason to believe that previously acquired knowledge is stored in the cerebral cortex, including the lateral temporal lobe, in those areas that originally process the information.)

Finally, Milner made the remarkable discovery that there seemed to be a kind of knowledge that H.M. could learn and remember perfectly

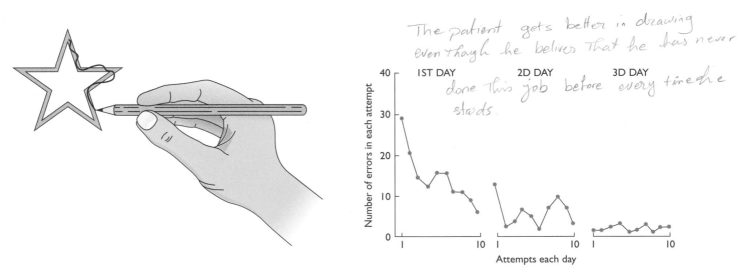

The patient gets better in drawing even though he belives That he has never done This job before every time he starts.

Patient H.M. learned successfully to trace between the two outlines of a star while viewing his hand in a mirror. He showed day-to-day improvement in this motor-skill task, yet on each day he had no recollection of having done the task before.

well—that is, a kind of memory that does not depend on the medial temporal lobe. In 1962, Milner unearthed evidence that H.M.'s defect in converting information from short-term memory to a long-term memory store was not absolute. In a famous experiment, Milner found that H.M. could learn to trace the outline of a star in a mirror, and his skill at it improved from day to day just as it would in a normal subject. Yet at the beginning of each day's test, he claimed that he had never done the task before.

These studies provided three fundamental insights into the biological nature of memory. First, Milner showed that lesions of medial temporal lobe structures, including the hippocampus, dissociate immediate memory from long-term memory, thereby validating on a biological level the fundamental distinction formulated by William James. Second, Milner disproved Lashley's idea of mass action. She found that restricted lesions of the medial temporal lobes had no effect on perception and intellectual functions but profoundly disrupted the ability to lay down new memories. Finally, she provided the first experimental evidence that there is more than one kind of memory.

TWO FORMS OF MEMORY STORAGE

Milner's finding that H.M. could learn and retain a mirror-drawing skill was initially interpreted to mean that motor skill learning has some special and separate neurological status. It was still thought that every other kind of learning is impaired. However, it subsequently came to be appreciated that motor skill learning is just one example of a large domain of learning and memory abilities, all of which are intact in H.M. and other amnesic patients. Furthermore, it became clear that the contrast between the kind of learning and memory lost in amnesic patients and the kind that is spared is not simply an effect of brain damage, but represents a fundamental distinction in the way all humans process and store information about the world. One kind of learning, the kind that is spared in amnesia, often has an automatic quality. Information, when retrieved, is not consciously experienced as a memory, as when we perform an acquired motor skill such as swinging a tennis racket. The learning is often accumulated slowly through repetition and is then expressed through performance but without evoking or requiring awareness of any past experience or even awareness that memory for the past is being used. The other kind of learning, the kind lost in amnesia, provides the conscious capacity to recollect past events.

As the psychologist Daniel Schacter and others have noted, philosophers and psychologists had introduced essentially this same distinction more than 100 years ago, on the basis of intuition and introspection. In his classic work *Principles of Psychology*, published in 1890, William James wrote separate chapters on habit (a mechanical, reflexive action) and memory (which involves conscious awareness of the past). Henri Bergson, the French philosopher, wrote in 1910 that the past can survive either as a bodily habit or as an independent recollection. In 1924, the psychologist William McDougall distinguished between *implicit* and *explicit* recognition, the former more automatic and reflexive, the latter involving conscious remembering of the past. Later, in 1949, the British philosopher Gilbert Ryle proposed the existence of two kinds of knowledge: one was concerned with *knowing how*, or knowledge of skills, and the other was concerned with *knowing that*, or knowledge of facts and events. Some years later the psychologist Jerome Bruner, one of the fathers of the cognitive revolution, called "knowing how" a *memory without record*. Memory without record reflects the way that encounters are converted into some process that changes the nature of the organism, its skills, or the rules by which it operates, even though they are virtually inaccessible as individual encounters. In contrast, he called "knowing that" a memory with record, a repository of information about the persons, places, and events of everyday life.

Indeed, a central feature of Freudian psychoanalytic theory, beginning late in the nineteenth century, was that experiences can leave their trace not only as ordinary conscious memories but also as essentially unconscious memories. These unconscious memories are inaccessible to awareness but can nevertheless have powerful effects on behavior. While these ideas were interesting, they did not on their own convince many scientists. What was finally needed was not philosophical debate but experimental inquiry into how the brain actually stores information. H.M.'s mirror-tracing of a star marked the beginning of a body of experimental work that would eventually establish the biological reality of two major forms of memory.

In 1968, Elizabeth Warrington and Lawrence Weiskrantz described a kind of test that amnesic patients could often perform normally. Instead of asking subjects to recall or recognize previously studied words, they presented the first few letters of the words as cues (e.g., MOT for MOTEL). The patients often responded to the cues by producing the previously studied word, even though they seemed to regard the test as a guessing game rather than a memory task. This phenomenon is now known as priming. The term refers to an improved ability to process, detect, or identify a stimulus, which is gained through having recently processed that stimulus.

Priming can be nicely illustrated in a test of picture naming. A subject is shown a picture of a particular airplane and asked to name it. The first time, the subject needs about 900 milliseconds (just under one second) to produce the word "airplane." Later, when the same airplane is presented again, the subject needs only about 800 milliseconds. Thus, following a single presentation of the airplane picture, the subject becomes

more capable of processing that specific object. This more efficient processing also occurs in amnesic patients who are unable to recognize the object as one they have seen recently. Examples of spared learning and memory abilities in amnesia soon multiplied to include, in addition to motor skill learning and priming, the learning of habits, classical conditioning, the learning of skills that have no motor component, such as the skill of learning to read mirror-reversed print, and many others.

There remains uncertainty about exactly how many distinct memory systems there are and how they should be named. Nevertheless, a consensus view has emerged about the major memory systems of the mind and about the brain areas that are most important for each memory system. The alternative classification schemes mentioned above simply use different terms for the same basic distinction. For example, memory for facts and memory for skills are known alternatively as memory with record and without record, explicit and implicit memory, or declarative and nondeclarative memory. For simplicity we will use one terminology. We refer to damage to the hippocampus and the medial temporal lobe, as in patient H.M., as affecting *declarative* memory, while we call the other forms of memory that remain intact *nondeclarative*. Declarative memory is memory for facts, ideas, and events—for information that can be brought to conscious recollection as a verbal proposition or as a visual image. This is the kind of memory one ordinarily means when using the term "memory": it is conscious memory for the name of a friend, last summer's vacation, this morning's conversation. Declarative memory can be studied in humans as well as in other animals.

Nondeclarative memory also results from experience but is expressed as a change in behavior, not as a recollection. Unlike declarative memory, nondeclarative memory is unconscious. Often, some recollective ability can accompany nondeclarative learning. We might learn a motor skill and then be able to remember some things about it. We might be able to picture ourselves performing it, for example. However, the ability to perform the skill itself seems to be independent of any conscious recollection. That ability is nondeclarative. Different forms of nondeclarative memory are thought to depend on different brain regions with names such as the amygdala, the cerebellum, the striatum, as well as on the specific sensory and motor systems recruited for reflexive tasks. Nondeclarative memory may be the only kind of memory available to invertebrate animals because they do not have the brain structures and brain organization that could support declarative memory. They do not, for example, have a hippocampus.

MECHANISMS OF MEMORY STORAGE: HOW ARE MEMORIES STORED?

What exactly changes in the brain when we learn and then remember? What ultimately happens in the brain depends on changes in the signalling between individual neurons, which in turn depends on the activities of molecules within the neurons. Declarative and nondeclarative memory recruit different brain systems and use different strategies for storing memory. Do these two distinct forms of memory utilize different molecular steps for storage, or are the storage mechanisms fundamentally similar? How does short-term storage differ from long-term storage? Do they occur at different loci, or can the same neuron store information for both short- and long-term memory?

The idea of studying molecular mechanisms of memory storage seems staggering, almost an impossibility. A mammal's brain is made up of an estimated 10^{11} nerve cells—one hundred billion—and the interconnections between these cells are many times more numerous. How are we to locate nerve cells critical for memory storage in this enormously large population? Fortunately, the task of identifying molecular mechanisms within cells can be simplified experimentally. Scientists can study forms of memory storage that involve only restricted portions of the vertebrate nervous system such as the isolated spinal cord, the cerebellum, the amygdala, or the hippocampus. Even more radically, scientists can study the simpler nervous systems of invertebrate animals. In the study of invertebrate animals it is sometimes possible to identify individual nerve cells that are directly involved in a particular kind of learning. One can then try to discover what molecular changes within these nerve cells are responsible for learning and memory storage.

By the end of the nineteenth century, biologists had come to appreciate that most mature nerve cells have lost their capacity to divide. As a result, we do not add an appreciable number of new nerve cells to our brain during our lifetime. This fact prompted the great Spanish neuroanatomist Santiago Ramón y Cajal to suggest that learning cannot cause the growth of new nerve cells. Rather, he proposed, learning might cause existing nerve cells to strengthen their connections with other nerve cells so as to be able to communicate more effectively with them. In order to store long-term memories, nerve cells might grow more branches and thereby form new or stronger connections. As a memory fades, the nerve cells might lose these branches and thereby weaken their connections. To take only the simplest example: A slight noise might cause you to startle a bit the first time you hear it. The noise activates pathways in the brain connected to the motor cells that control your muscles. But if that noise is repeated several times over an extended period, those connections might be weakened so that you now no longer startle in response to that noise.

Ramón y Cajal's suggestions about memory mechanisms were interesting and influential, but, as with the early suggestions about multiple memory systems, the mere suggestion of a possible mechanism was not sufficient. What was needed were simple nervous systems for study, so that one could examine nerve connections while an animal learned. Only in that way could one determine whether changes in the strength of nerve connections underlie memory storage. Over the last 40 years, scientists have developed a number of model systems for the specific purpose of studying mechanisms that might contribute to memory storage, with the ultimate goal of identifying its cellular and molecular basis.

This approach to memory storage began with cell biological studies of the marine snail *Aplysia,* a simple invertebrate animal, and was soon followed by genetic studies of the fruit fly *Drosophila.* The idea was that simple animals have simple brains, so their behavior as well as their ability to learn and remember is accessible to cellular and molecular analysis. As scientists gained confidence in these approaches, they have extended them to mice, taking advantage of new techniques for making changes in individual genes in the brains of mice and exploring their effects on memory storage.

The sea hare (Aplysia californica). *The relatively simple nervous system of this animal makes it amenable to cellular and molecular studies of learning and memory.*

SIMPLE SYSTEMS FOR CELLULAR AND MOLECULAR STUDIES

In contrast to the brain of a mammal with its 100 billion nerve cells, the central nervous system of a simple invertebrate animal like *Aplysia* has approximately 20,000 nerve cells. In *Aplysia* these cells are clustered into groupings called ganglia, each containing about 2000 cells. A single ganglion such as the abdominal ganglion contributes not to one behavior but to a set of different behaviors: gill and siphon movements, control of heart rate and respiration, the release of ink (a defensive response), and the release of reproductive hormones. Thus, the number of nerve cells contributing to the simplest behavioral acts— acts that can nevertheless be modified by learning—may be as few as a hundred.

One great advantage of *Aplysia* and other invertebrate animals for cellular studies is that many of these nerve cells are distinctive and can be identified as unique cells from animal to animal. Indeed, some of the nerve cells are nearly one millimeter in diameter, so large that they can be recognized with the naked eye without the aid of a microscope. As a result, the scientist can identify many of the cells involved in a simple

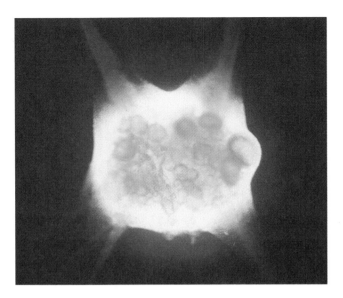

Aplysia ganglion. *Each of the 10 ganglia in* Aplysia *contains about 2000 nerve cells. Some of these cells are large enough (1 millimeter in diameter) to be seen with the naked eye.*

SIMPLE SYSTEMS FOR GENETIC STUDIES

The cell biological studies we have just described were soon complemented by genetic studies. It had long been obvious to breeders of domestic animals that many physical characteristics are inherited, including the shape of the body, the color of the eyes, and even temperament and physical strength. If even temperament can be inherited through the action of genes, the question naturally arises: Are more subtle components of behavior in any way determined by genes? If so, do genes have a role in the modification of behavior? Do they have a role in learning and memory storage? The idea arose that scientists might be able to identify specific genes important for learning and for memory storage. Identifying important genes could then lead to discovery of their products, the proteins that play important roles in

behavior, then construct a "wiring diagram" showing how these cells are linked together. One can then ask what happens to specific neurons within the behavioral circuit when the animal learns.

Even simple animals such as *Aplysia* exhibit several different kinds of learning, and each kind gives rise to both short-term memory lasting minutes and long-term memory lasting weeks, depending on the number and spacing of the training trials. For example, *Aplysia* is capable of habituation—the ability to learn to ignore a benign stimulus that is trivial and not informative—as well as sensitization—the ability to learn to modify its behavior when a stimulus is aversive. Finally, *Aplysia* can learn classical and operant conditioning—it can learn to associate two stimuli or a stimulus and a response. It therefore became possible to explore the cellular mechanisms contributing to different forms of learning and memory storage in these animals and to identify specific molecules that are critical for short- and long-term memory.

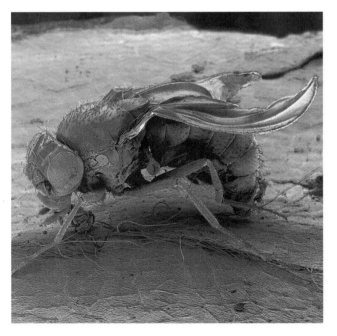

The fruit fly (Drosophila melanogaster) *has been useful for genetic studies of learning and memory.*

cellular function, and ultimately could lead to an unraveling of the molecular steps involved in making and storing memory.

Gregor Mendel, the father of genetics, carried out his work on plants—on peas and their pods. It was the American biologist Thomas Hunt Morgan at Columbia University who turned the study of genetics to experimental animals. Early in this century, Morgan saw the potential of the fruit fly *Drosophila* to serve as a model organism for genetic studies. Morgan appreciated that *Drosophila* has only 4 pairs of chromosomes in its germ cells, as compared with 7 pairs of chromosomes in Mendel's peas, 17 in *Aplysia*, and 23 in humans. This small fly can be reared by the thousands in the laboratory. It is possible to produce mutations in single genes by chemical means and, because of the fly's relatively short generation time of two weeks, to quickly breed many flies carrying the mutant gene.

The critical step for studying the genetics of behavior, learning, and memory in *Drosophila*

was taken in 1967 by Seymour Benzer at the California Institute of Technology. Using chemical techniques to produce mutations in single genes, Benzer began to examine the effects on behavior of changing one gene at a time. Having first identified a number of fascinating mutants that affected courtship, visual perception, and circadian rhythms, Benzer turned this genetic approach to the problem of learning and memory storage. From mutants with memory defects, Benzer was able to identify several proteins that are important for nondeclarative forms of memory storage. It was immediately evident that some of these proteins were the same as those identified independently in molecular biological studies of nondeclarative memory in *Aplysia*.

COMPLEX SYSTEMS FOR GENETIC STUDIES

What about declarative forms of memory storage? What molecules are used for this type of memory? Although experimental animals cannot *declare* anything, they can learn and remember in ways that have many of the critical features of declarative memory. However, for a long time it was not feasible to apply to declarative memory the sort of cell and molecular biological analysis that had become routine in *Aplysia* and *Drosophila*. This situation changed dramatically in 1990 when Mario Capecchi at the University of Utah and Oliver Smythies at the University of Toronto developed methods for *gene knockout* in mice. This technique made it possible to eliminate specific genes in the mouse genome and study the effects of their absence. A few years earlier, Ralph Brinster at the University of Pennsylvania and others had developed methods for introducing and activating new genes—genes that are not normally present in the mouse or not normally strongly expressed. As a result of these two

Seymour Benzer, the American biologist who pioneered the genetic analysis of behavior and learning in Drosophila.

human genome project. Now that it is possible to also carry out genetic studies in mice, molecular biological studies of learning and memory have the opportunity to achieve their full promise.

FROM MOLECULES TO MIND: THE NEW SYNTHESIS

As we shall see in later chapters, molecular biological approaches have joined with those of systems neuroscience and with cognitive psychology to form a common, unified science that has proven to be as fascinating from a molecular point of view as it has from a behavioral one. The growing partnership between these once independent fields of study is leading to a new synthesis of knowledge about memory and the brain.

The Barnes maze is a tool used to study learning and memory in genetically altered mice. Mice and rats prefer dark enclosures to open, well-lit areas. An animal placed on the Barnes maze learns the location of the one hole that permits escape from the brightly lit table surface.

advances, biologists can now alter any gene in the mouse and determine how such a change affects the functioning of nerve cells in the hippocampus or other areas important to memory. They can also determine how such a change affects declarative memory in the intact behaving animal.

These advances launched the modern molecular study of declarative memory in mammals. Indeed, the mouse already had many advantages for the study of memory, including its mammalian heritage and its neuroanatomical, physiological, and genetic closeness to humans. Moreover, the mouse genome is being mapped in parallel with the human genome as part of the

On the one hand, interesting new molecular properties of nerve cells, and especially their connections, are revealed time and time again during studies of learning. These molecular findings are pointing the way to explanations of how nerve connections change during learning and how those changes are maintained over time as memory. On the other hand, systems neuroscience and cognitive science are explaining how nerve cells work together in neural circuits, how learning processes and memory systems are organized, and how they operate. In addition, investigations of brain systems and behavior are providing a road map for molecular studies, a map that identifies components of memory and the areas of the brain where these components can be studied in detail. Indeed, many of the molecular insights have come to light only because nerve cells located in a particular neural circuit could be examined with a particular form of memory in mind. Thus, the study of memory infuses cell and molecular biology with a new fascination—a fascination stemming from the possibility of investigating the biology of important mental processes.

From a cognitive point of view, cellular and molecular approaches have made headway toward answering some of the key unsolved problems in the psychology of memory. What is the molecular relationship between nondeclarative and declarative memory storage? How do short-term forms of memory relate to long-term forms? Most important, molecular approaches have provided an initial bridge connecting behavior in intact animals to molecular mechanisms in single cells. Thus, what were formerly only psychological constructs like association, learning, storing, remembering, and forgetting can now be approached in terms of cell and molecular mechanisms and in terms of brain circuits and brain systems. In this way, it has become possible to achieve deep insights into fundamental questions about learning and memory.

In the chapters that follow, we describe what is now known about the cognitive psychology of learning and memory and its biological underpinnings.

We take a close look at the cellular and molecular events that occur within neurons, especially in the simple forms of learning that can be studied in animals like *Aplysia* and *Drosophila*. These studies have revealed a molecular switch, shared by both declarative and nondeclarative memory, that converts short-term into long-term memory. Additional insights into the cellular biology of memory have come from direct studies of tissue taken from areas of the vertebrate brain important for declarative memory. In these brain areas, the connection strength between neurons can change quickly and last for a long time, a phenomenon known as long-term potentiation (LTP). A major theme in these chapters is that long-term memory involves changes in the structure of nerve cells. Depending upon the type of learning, the nerve cells involved can make either more and stronger connections or fewer and weaker connections.

In other chapters, we look at what the experimental studies of animals and humans tell us about the nature of memory and the organization of the brain systems that support memory. These studies have taught us about the power of memory and its imperfections, about the factors that influence the strength and persistence of memory, and about the important contribution that forgetting makes to normal memory function. These studies have also identified the brain systems that support declarative memory and shown how they operate. Finally, this work has revealed an unexpected variety of unconscious, nondeclarative types of memory and has identified the brain systems important for each type. These kinds of memory carry the traces of past experience and exert strong influences on behavior and mental life, but they can operate outside of awareness and do not require any conscious memory content.

By uniting the perspective of cellular and molecular biology and the perspective of neural systems and cognitive psychology, we hope to convey the important advances that have occurred, as well as the new synthesis that is beginning to be achieved in understanding how memory works.

Robert Rauschenberg, Reservoir *(1961). By culling images and materials from the world around him, Rauschenberg (1925–) blurred the distinction between art forms and everyday experience. Employing a contemporary fresco process, Rauschenberg here uses clocks as if to indicate the connection that time provides in linking past and present memories.*

2 | Modifiable Synapses for Nondeclarative Memory

In 1957, when Brenda Milner first described a catastrophic memory loss in patient H.M., she and other scientists supposed that this memory loss applied to all areas of knowledge. As we saw in Chapter 1, in 1962 Milner made the equally surprising discovery that H.M. could learn some new things. Specifically, he could learn new motor skills. When H.M. tracked a moving target, or filled in the outlines of a star by looking at it only in a mirror, his performance improved progressively, exactly like that of normal subjects. However, there was one important difference between H.M.'s performance and that of a normal person: in each case H.M. was not in any way aware that he had ever done the task before.

For a number of years Milner and other memory researchers thought that people with brain damage similar to H.M.'s retained only one specialized and restricted type of long-term memory—they could still learn and remember motor skills. Over the course of the next two decades, however, it became clear that motor skills represented just the tip of the iceberg. Larry Squire in San Diego and others conducted further studies of patients who, like H.M., had bilateral lesions of the medial temporal lobe. They found that these patients retained a large repertory of memory capabilities, which we now call nondeclarative memory. All of these memory capabilities shared the remarkable feature that they are normally inaccessible to the conscious mind. The recall of these kinds of memory is completely unconscious.

What we now call nondeclarative memory includes a large family of different memory abilities sharing one feature in particular. In each case, memory is reflected in performance—in how we do something. This kind of memory includes various motor and perceptual skills, habits, and emotional learning, as well as elementary reflexive forms of learning such as habituation, sensitization, and classical and operant conditioning. Thus, nondeclarative memory typically involves knowledge that is *reflexive* rather than *reflective* in nature. For example, when you first learned to ride a bicycle, you probably paid a great deal of conscious attention to steering the front wheel with the handle bars and you also focused on pushing the pedals, first with the left foot and then with the right. But once bicycle riding becomes well practiced, the skill is stored as nondeclarative memory. You continue to watch the road attentively, but now you steer and pedal automatically—reflexively, not reflectively. You do not attempt to recall consciously that you now need to push down with your right foot, then with your left foot. If you *do* pay attention to all these movements, you may fall off the bicycle. Similarly, when playing tennis you naturally bring your racquet head up in front of you when hitting a high forehand volley and you bring the racquet head down when you hit a low forehand ground stroke. Once you are practiced at these motions, you do not rehearse them to yourself before carrying them out.

Scientists were excited to have uncovered a large category of knowledge that operates in parallel with declarative forms of knowledge. Yet the discovery of nondeclarative memory as a distinct form of memory is also interesting for two larger reasons. First, it provides biological evidence that unconscious mental processes actually exist. That some memory processes are unconscious was first suggested by Sigmund Freud, the founder of psychoanalysis and the discoverer of *the* unconscious. But what is fascinating about nondeclarative memory is that it bears only superficial resemblance to *the* Freudian unconscious. Nondeclarative knowledge is unconscious, but it is not related to conflict or to sexual strivings. Moreover, even though you successfully perform the tasks encoded in nondeclarative memory, the encoded information will not enter your consciousness. Once stored in nondeclarative memory, *this* unconscious never becomes conscious.

Second, it turned out that many years earlier the behavioral psychologists had characterized a number of nondeclarative forms of memory. In fact, because these forms are so amenable to experimental manipulation, the behaviorists had made the study of learning that gives rise to these forms of memory their central concern.

At the beginning of the twentieth century, two major nondeclarative learning procedures—nonassociative and associative learning—were delineated by the Russian physiologist Ivan Pavlov, by the American psychologist Edward Thorndike, and by others. Habituation and sensitization are examples of *nonassociative* learning. In these types of learning, a subject learns about the properties of a *single* stimulus—such as a loud noise—by being exposed to it repeatedly. Classical and operant conditioning are examples of *associative* learning. Here a subject learns about the relationship between two stimuli (*classical conditioning*) or about the relationship of a

Ivan Pavlov (1849–1936), the Russian physiologist who discovered classical conditioning. Studies of classical conditioning in turn led to the discovery of habituation and sensitization.

acquisition of knowledge rather than its retention, they failed to appreciate, and did not really care, that the retention of nondeclarative knowledge is unconscious. And, in treating nondeclarative knowledge as if it explained the acquisition of *all* knowledge, the behaviorists largely ignored what we now call declarative memory.

Some years after Milner discovered that the amnesic patient H.M. could learn simple motor tasks, other researchers found that amnesic patients also have perfectly normal memory for simple associative learning. Thus, rather than representing all of learning, the studies of elementary forms of learning, such as learning to expect food when you see a green light appear, represented a special case—a type of learning that specifically gives rise to performance without awareness.

Despite this limitation, the work of the behaviorists proved very valuable, for in the course

Edward Thorndike (1874–1949), the American psychologist at Columbia University who discovered instrumental conditioning, or trial-and-error learning, now often called operant conditioning.

stimulus to the subject's behavior *(instrumental* or *operant conditioning).* Thus, in classical conditioning, an animal that learns to associate a bell with the taste of food will salivate when it hears the bell. In operant conditioning the animal will learn to associate pressing a bar or a key with the delivery of food: when it presses the bar, it expects to receive something to eat.

The behavioral psychologists focused on these forms of learning during the first half of this century because these psychologists insisted that the acquisition of skills and knowledge be studied objectively. But, by focusing so extensively on the

of their work, the behaviorists showed that the rules that govern simple forms of nondeclarative memory are very general and apply not only to humans but also to experimental animals, even very simple ones.

In this chapter, we shall focus on the simplest case of nondeclarative memory, *habituation*. Nondeclarative memory has been studied most effectively in the simple reflex systems of invertebrates and vertebrates. The cell biological insights that have been obtained from these simple systems, however, have proven to be true for more complex animals and more complex forms of memory.

THE SIMPLEST CASE OF NONDECLARATIVE MEMORY: HABITUATION

When we hear a sudden noise such as a toy gun being fired behind us, a number of autonomic changes are triggered in our bodies. Our hearts beat quicker and our breathing becomes rapid; our pupils dilate; and our mouths may feel dry. If the noise is repeated, however, these responses will abate. This is habituation, a type of learning experienced routinely in one form or another by all of us and also by the simplest of animals. Thus, one can become accustomed to initially distracting sounds and learn to work effectively in an otherwise noisy environment. One becomes habituated to the sound of a clock in the study, and to one's own heartbeat, stomach movements, and to the clothes one wears. These enter awareness rarely and only under special circumstances. In this sense, habituation is learning to recognize, and ignore as familiar, unimportant stimuli that are monotonously repetitive. Thus, city dwellers may scarcely notice the noise of traffic at home but may be awakened by the chirping of crickets in the country.

Habituation also operates to eliminate inappropriate or exaggerated defensive responses. This is beautifully illustrated in the following fable from Aesop:

> A fox, who had never seen a turtle, when he fell in with him for the first time in the forest was so frightened that he was near dying with fear. On his meeting with him for the second time, he was still much alarmed but not to the same extent as at first. On seeing him the third time, he was so increased in boldness that he went up to him and commenced a familiar conversation with him.

As a result of habituation to common, innocuous stimuli, an animal can learn to ignore a large number of stimuli that are not crucial to its survival. The animal can instead focus on stimuli that are novel or stimuli that signal satisfying or alarming consequences. Habituation is an essential feature of animal training, by which means undesired responses, such as gun-shyness in dogs or skittishness to car sounds in horses, can be eliminated. Thus, a police dog may be afraid at the sight of a gun, having associated it with loud noises. But after being repeatedly exposed to the gun in a policeman's holster, the dog learns that, by itself, a gun is harmless and remains so until the gun is drawn.

Habituation is not restricted to escape responses. It may also lessen the frequency with which sexual responses are expressed. Given free access to a receptive female, a male rat will copulate six or seven times in a period of one or two hours. After the last copulation he appears exhausted and becomes inactive for 30 minutes or longer. This is sexual habituation, not fatigue. An apparently exhausted male will promptly resume mating if a new female is made available. Similarly, a pair of male and female rhesus monkeys first placed together in a cage will couple rapidly and often, with little foreplay. After several days the frequency of copulation decreases and every act is preceded by more lengthy preliminaries, each partner examining and stimulating the other. However, if the male is exposed to a new female partner, he becomes immediately aroused and potent, and the preliminaries are dropped.

The fact that after repeated exposure a dog reacts differently to a gun in a holster than to a gun in the hand suggests that the animal has learned and somehow remembered the appearance of the gun in the two different contexts. It is useful to think of the dog as building an internal representation in its brain of the gun in its holster, a representation sufficiently rich in detail that the dog can recognize the situation on subsequent occasions. Thus, habituation reveals something about the organization of perception itself.

In fact, we have learned much from the study of habituation about how internal representations develop in the brain. For example, developmental psychologists make use of habituation to study perception and cognition in human newborns.

Briefly, the procedure involves habituating the newborn infant to one stimulus, say a blue square, and then examining how it responds to a novel stimulus, say a red square. In a typical experiment, a six-month-old infant is briefly shown a blue square. On seeing that stimulus for the first time, the infant's eyes will focus intently on the stimulus and its heartbeat and respiratory rate decrease. With repeated presentations of the blue square, the responses habituate. If the infant is now instead shown a red square, its visual attention will immediately focus on the stimulus and its heartbeat and respiratory rate will again decrease, indicating that the infant can distinguish the novel red square from the familiar blue one. The infant has revealed its perceptual capacity to distinguish blue from red. Using this procedure, we have learned that infants are able to categorize colors and aspects of speech in ways that adults do.

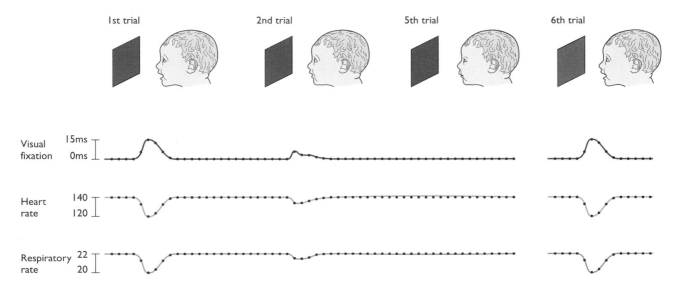

Habituation can be used to study perception in newborn infants. When an infant is first shown a novel blue square, its visual attention is captured by the stimulus and its heart rate and respiration decrease. As the blue square is shown repeatedly, the infant learns to ignore the familiar stimulus and its responses habituate. However, when a novel red square is shown to the infant, the new stimulus will immediately recapture the infant's visual attention, and its heart and respiratory rate will again decrease. In this way scientists have determined that an infant can distinguish one color from another.

Studies of habituation in simple experimental animals provided the first evidence of how learning and memory storage take place in the brain. These studies showed that learning modifies the capability of nerve cells (neurons) to signal each other. A memory is formed when these modifications are made to persist. Before exploring memory formation further, however, we first must consider the relation of individual nerve cells to the brain as a whole, starting with the neuron doctrine of Santiago Ramón y Cajal.

NEURONS: SIGNALING ELEMENTS OF THE BRAIN

The nerve cells that make up the brain are signaling devices of a rather remarkable kind. Their signaling capabilities underlie all aspects of our mental life, from sensory perception to the control of movement, from the generation of thought to the expression of feeling. Understanding the signaling properties of neurons therefore is essential for understanding the biological basis of any aspect of behavior.

Our initial insights into how signaling occurs in the brain date from the beginning of the twentieth century and especially from the extraordinary contributions of Santiago Ramón y Cajal, the great Spanish anatomist. Based on his studies, Ramón y Cajal formulated the "neuron doctrine," which states that the brain is made up of discrete cells, called nerve cells or neurons, each delimited by an external membrane. He proposed that these neurons are the elementary signaling units of the brain. For these studies Ramón y Cajal was awarded the Nobel Prize in physiology or medicine in 1906.

Ramón y Cajal pointed out that the nerve cells in all animals have surprisingly similar anatomical features. From this finding, we now

Santiago Ramón y Cajal (1852–1934), the Spanish neuroanatomist who appreciated that an exact knowledge of the brain was essential for developing a rigorous understanding of mental processes. Ramón y Cajal developed the neuron doctrine, the idea that the nerve cell is the signaling unit of the brain. In so doing, he provided the key anatomical evidence that one nerve communicates with another through specialized contacts, now called synapses.

know that the different learning capabilities of different animals are related not so much to the *types* of nerve cells that an animal has in its brain as to the number of nerve cells and the way they are interconnected. With some exceptions, the greater the number of nerve cells, and the more complex their patterns of interconnection, the greater an animal's capability for different types of learning. Some invertebrate animals, such as snails, have 20,000, or 2×10^4, neurons in the

Different learning ability depends on # of neurons and how they are interconnected. not the type of neurons.

brain. A fruit fly has about 300,000 nerve cells (3×10^5). By contrast, a mammal such as a mouse or a person has from 10 billion to 100 billion nerve cells (10^{10} to 10^{11}). Each neuron in the brain in turn makes about 1000 connections to other neurons, at specialized connecting junctions called *synapses*. This means that in the human brain there are a total of about 10^{14} synaptic connections. One of the insights of the modern biology of memory is the finding that the individual connection made between two neurons is an elementary unit of memory storage. Thus, the 10^{14} connections in the human brain provide one rough indicator of our maximal memory storage capacity.

Signaling in nerve cells often is initiated by physical events in the environment that impinge on our bodies—mechanical contact, odorants, light, or pressure waves. The surprising fact about neuronal signals is that they are all remarkably stereotyped. Nerve signals that convey visual information are identical to those that carry information about sounds or odors. In turn, the incoming signals that carry sensory information into the nervous system are similar to the outgoing signals that convey the commands for movement. Thus, one of the key principles of brain function is that the nature of the information conveyed by a nerve signal is determined not by the nature of the signal, but by the particular pathways that the signal travels in the brain. The brain analyzes and interprets *patterns* of incoming electrical signals along specific and dedicated pathways and in this way processes sight in one set of pathways and sound in another set of pathways. We see the face of a person rather than hear his or her voice because the nerve cells in the retinas of our eyes connect to those parts of the brain (the visual system) that process and interpret visual information, information about sights.

Ramón y Cajal and his contemporaries discovered that every neuron has four components:

a cell body, a number of dendrites, an axon, and a family of axon terminations, called the presynaptic terminals. The cell body is the large globular central portion of the neuron, containing the nucleus, which in turn houses the DNA that encodes the neuron's genes. Surrounding the nucleus is the cytoplasm, the cell sap of the cell body, which contains a variety of molecular machinery for synthesizing and packaging proteins necessary for the cell to function. The cell body gives rise to two types of long slender threads, or extensions, generically known as *nerve cell processes*: the dendrites and the axon. The *dendrites* typically consist of elaborately branching processes that extend from the cell body, often in the form of a tree, and form the input component or receptive area for incoming signals. The *axon,* the output component of the neurons, is a tubular process that extends from the cell body. Depending on the cell's specific function, the axon can travel distances as short as 0.1 millimeter to as long as a meter or more. Near its ending the axon divides into many fine branches, each of which has a specialized terminal region called the *presynaptic terminal*. The presynaptic terminals contact the specialized receptive surfaces of other cells, often located on dendrites. Through this contact at the synapse, a nerve cell transmits information about its activity to other neurons or to organs such as muscles or glands.

What distinguished Ramón y Cajal from his contemporaries was that he was able to go beyond anatomical description. He had the uncanny ability to look at static structure—at an anatomical section of a set of neurons under a microscope—and obtain insight about function. For example, he had the remarkable insight that these four anatomical components of the neuron had distinctive roles in signaling. Based on this insight, he formulated the idea that neurons are *dynamically polarized* so that information flows in a predictable and consistent direction within

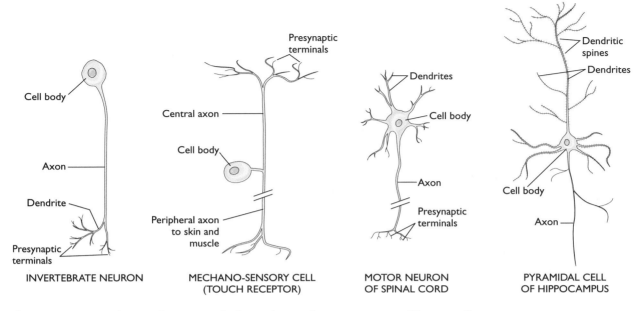

There are a variety of types of neuron in the brain, but, with rare exception, all have a cell body, dendrites, axon, and presynaptic terminals. As we shall see in the figure on page 32, the axon ends in a number of presynaptic terminals.

each nerve cell. Information is received at the dendrites and the cell body, and from these receiving sites the information is conveyed to the axon, and from the axon to the presynaptic terminals. Subsequent work has proven him right. In the decades between 1920 and 1950 it was found that neurons use not simply one, but two types of signals: (1) They use all-or-none, stereotypic *action potentials* for signaling within the neuron—that is, for passing information from one *region* or compartment of the neuron to the other, such as from dendrites to the cell body and from the cell body to the axon and its presynaptic terminal. (2) They use graded *synaptic potentials* to pass information from one *nerve cell* to another through the process of *synaptic trans-*

mission. Both types of signals, as we shall see, are important for memory storage.

NEURAL SIGNALS

Before turning to consider the action potential and the synaptic potential, we need to consider the *resting potential*, the baseline condition upon which all other cellular signals are expressed. The external plasma membrane of the cell maintains at rest an electrical difference of about 65 millivolts; this is the resting potential. It results from an unequal distribution of sodium, potassium, and other ions across the nerve cell membrane such that the inside of the cell membrane is negatively charged in relation to the outside. Because the outside of the membrane is arbitrarily defined as zero, we say the resting membrane potential is minus 65 millivolts (-65 mV).

Action potentials and synaptic potentials result from perturbations of the membrane that cause the membrane potential to increase or decrease with respect to the resting membrane potential. An increase in membrane potential from, say, -65 to -75 millivolts is called *hyperpolarization*; a reduction of the membrane potential, from -65 to -50 millivolts, is called *depolarization*. As we shall see later, depolarization increases a cell's ability to generate an action potential and is therefore excitatory. Conversely, hyperpolarization makes it less likely that a cell will generate an action potential and is therefore inhibitory.

The action potential is a depolarizing electrical signal that travels from the dendrites and cell body of the neuron, along the entire length of the axon, to its presynaptic terminals, where the neuron contacts another nerve cell. Action potentials are so named because they are signals that propagate *actively* along the axon. The precise mechanism of the electrical signal can be a bit tricky to visualize, but you don't need to understand it in detail to follow the discussions in this book.

The action potential is a change in the electric potential across the external membrane of the cell created by the movement of sodium ions (Na^+) into the cell and the subsequent movement of potassium ions (K^+) out of the cell through specific pores in the cell membrane called *ion channels*. The ion channels open and close in a precise sequence along the path of the signal, creating a change in the potential that travels through the cell. Action potentials move along the membrane of the axon without failure or distortion, at conduction velocities that range between 1 and 100 meters per second. The action potential is a rapid and transient, all-or-none electrical signal with an amplitude of 100 to 120 millivolts and a duration

at any one spot of from 1 to 10 milliseconds. The amplitude of the action potential remains constant throughout the axon because the all-or-none impulse is continually regenerated by the membrane as it travels down the axon.

Ramón y Cajal appreciated that neurons communicate with one another at highly specialized contact points called synapses. Some of the most remarkable activities of the brain, such as learning and memory, emerge from the signaling properties of synapses, so we shall focus on them in some detail. Whereas the signal in the axon—the action potential—is a large, invariant all-or-none signal, the signal at the synapse—the synaptic potential—is graded and modifiable.

A typical synapse has three components: a presynaptic terminal, a postsynaptic target cell, and a small space in between these two processes, separating the two neurons. This space, called the *synaptic cleft*, is about 20 nanometers (2×10^{-8} meters) wide. The presynaptic terminal of one cell communicates across the synaptic cleft with either the cell body or the dendrites of the postsynaptic target cell.

The current produced by the action potential in the presynaptic cell cannot jump directly across the synaptic cleft to activate the postsynaptic target cell. Instead, the signal undergoes a major transformation at the synapse. As the action potential reaches the presynaptic terminal, the electrical signal leads to the release of a simple chemical substance called a *chemical synaptic transmitter* or *neurotransmitter*. This substance spills into the synaptic cleft, where it acts as a signal to the target cell. As we shall see below, once the neurotransmitter has diffused across the synaptic cleft it is recognized by and binds to receptor molecules on the surface of the postsynaptic cell. The common neurotransmitters used by nerve cells are either amino acids or their derivatives such as glutamate, gamma-aminobutyric acid (GABA), acetylcholine, epinephrine, norepinephrine, serotonin, and dopamine.

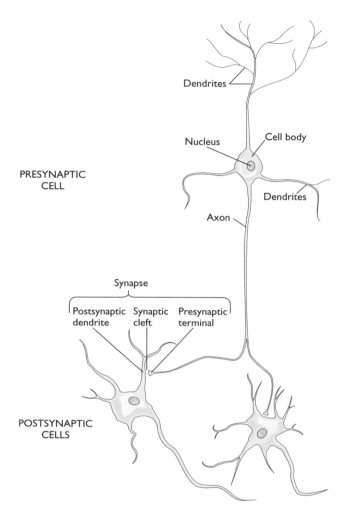

PRESYNAPTIC
CELL

POSTSYNAPTIC
CELLS

The long, thin axon of a neuron branches at its end into numerous presynaptic terminals that form synapses with the dendrites of one or more postsynaptic cells. The terminals of a single axon may synapse with as many as 1000 other neurons. The axons of many vertebrate neurons are covered with a fatty sheath called myelin, which acts to speed conduction. For simplicity, the myelin sheaths have been omitted from this and all other drawings.

ferent tissue types that became specialized into different functional systems such as the heart and circulatory system and the stomach and the digestive system. In turn, not one, but two kinds of chemical signals evolved to coordinate the activities of the various tissues: hormones and synaptic transmitters.

These two forms of chemical communication share certain features in common. In hormonal action, a gland cell releases a chemical messenger (a hormone) into the bloodstream to signal a distant tissue. For example, after a meal the level of the sugar glucose in the blood rises. This increase in glucose signals certain cells in the pancreas to release the hormone insulin, which acts on insulin receptors in muscle so that glucose is taken up into muscle cells, then converted and stored as glycogen, the energy reserve form of glucose. In synaptic transmission, a neuron releases a chemi-

Sir Bernard Katz (1911–), the British neurophysiologist who pioneered the modern analysis of synaptic transmission. He discovered that chemical synaptic transmitters are released not as single molecules but in multimolecular packets containing about 5000 molecules. Each packet is called a quantum and is contained in an organelle called the synaptic vesicle.

Chemical signaling is not limited to synaptic transmitters or to nerve cells of the brain. Rather, it is a universal mechanism of communication used by all cells in all multicellular organisms. As multicellular organisms began to appear hundreds of millions of years ago, they evolved dif-

cal messenger (a synaptic transmitter) to signal an adjacent target cell.

There are, however, two key differences between hormones and synaptic transmitters. The first is that synaptic transmitters typically operate over a much shorter distance than do hormones. What makes synaptic transmission special is that the membrane of the cell receiving the signal lies very close to the cell releasing the signal. As a result, synaptic transmission is much faster than hormonal signaling and far more selective in its targets. As we shall see, the close apposition of the two cells is central to the neuron's ability to store very specific information of the sort required for memory. The second difference between hormones and synaptic transmitters, which we shall describe in more detail later, is that a single synaptic transmitter is able to produce a variety of different responses in the target cell. By

contrast, hormones tend to act in the same way in a given set of target cells.

Biologists had appreciated several of these distinctive features of synaptic transmission by the 1930s. But their ideas were put on a new scientific footing in the 1950s and 1960s by the work of Sir Bernard Katz, a neurophysiologist at University College in London, England, who worked out many of the details of how synaptic transmission proceeds. For example, Katz and his colleagues found that as the action potential invades the presynaptic terminals, it opens up membrane channels for calcium ions (Ca^{2+}) that cause a large and rapid increase in Ca^{2+} into the presynaptic terminals. It is this large and rapid increase in Ca^{2+} that leads to the release of the chemical transmitter. The transmitter then diffuses across the synaptic cleft toward the postsynaptic cell. Finally, interaction of the neurotransmitter with

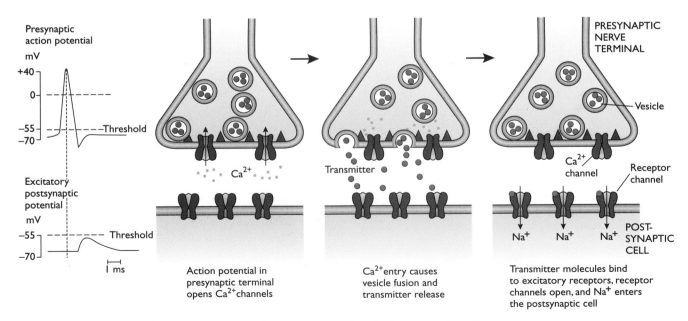

A chemical signal passes from a presynaptic cell to a postsynaptic cell. A presynaptic action potential triggers the release from a synaptic vesicle of a quantum of chemical transmitter into the synaptic cleft. The binding of transmitter molecules to postsynaptic receptors sets off a sequence of steps that results in the creation of an excitatory (or inhibitory) postsynaptic potential.

receptors in the postsynaptic cell causes a depolar-izing excitatory synaptic potential in the post-synaptic cell, which, if large enough, may generate an action potential in that cell.

The synaptic potential is an electrical signal, as is the action potential. Nevertheless, the two differ markedly. Whereas the action potential is typically a large signal of about 110 millivolts, the synaptic potential is a much smaller signal of anywhere from a fraction of a millivolt to several tens of millivolts, depending on a number of fac-tors, including how many presynaptic terminals are active and have released neurotransmitter that reaches the same postsynaptic cell. In addi-tion, the action potential is all-or-none; the synaptic potential is graded in strength. Finally, the action potential propagates actively and with-out fail from one end of the neuron to the other. The synaptic potential propagates passively and dies out unless it triggers an action potential.

One of Katz's dramatic discoveries was his finding that synaptic transmitters are released not as single molecules but as one or more packets of fixed size, each packet containing about 5000 molecules. Each of these packets is released in an all-or-none fashion. Katz called these packets *quanta* and appreciated that they were the ele-mentary units of chemical transmitter release.

NEW INSIGHTS FROM THE ELECTRON MICROSCOPE

Katz made his discovery that synaptic transmitter is released in packets in the 1950s, just at the time that researchers began using the electron micro-scope to observe nerve cells and, as a result, just when the first high-resolution pictures of the sub-cellular structure of the neuron became available. These pictures revealed that nerve cells shared with other cells of the body various well-defined subcel-lular structures called *organelles*. These organelles include the nucleus, which contains the genes of the cell, and the endoplasmic reticulum, which manu-factures protein. Each of these organelles is sur-rounded by a membrane similar to the one that makes up the external surface of the cell.

In addition to the organelles common to all cells, these pictures also revealed structures unique to nerve cells. Most prominent was a cluster of small round images—tiny spheres, or *vesicles*—about 50 nanometers in diameter. Because these vesicles were clustered in the presynaptic terminal, they suggested to Katz that they were *synaptic vesicles* and that they stored the packets of 5000 molecules that constitute a quantum of chemical synaptic transmitter and therefore served as the *structural units* of quantal release.

At the time Katz made these observations, it was already well known that the external surface membrane of all cells has a mechanism called *ex-ocytosis* for expelling bulk quantities of various substances out of the cell. Katz proposed that the presynaptic nerve terminal releases packets of transmitter from the synaptic vesicles by means of exocytosis. Soon after this suggestion was

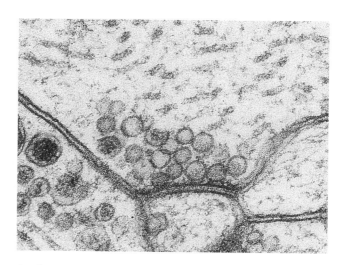

An electron micrograph of a synapse. Numerous vesicles, each storing a single quantum of synaptic transmitter, are clustered near the center. They are ready to be released at the active zone, the dark area along the presynaptic side of the synaptic cleft that serves as the docking, fusion, and re-lease site for vesicles.

made, the French anatomist René Couteaux was able to confirm it. He found that synaptic vesicles fuse and coalesce with the external surface membrane of the presynaptic terminal and release their content—the entire 5000 molecules—in an all-or-none fashion by means of exocytosis into the synaptic cleft. Couteaux further found that the synaptic vesicles do not fuse and undergo exocytosis at any arbitrary point in the presynaptic terminal but only at specialized and restricted sites that he called *active zones*. It is at these active zones that the Ca^{2+} channels are located that allow Ca^{2+} to enter the presynaptic terminals. Vesicles are normally released spontaneously at the active zone at very slow rates, even in the absence of an action potential. The rate of vesicle release is greatly increased by the Ca^{2+} influx that occurs with each action potential.

Once released into the synaptic cleft, the 5000 molecules of chemical transmitter diffuse to the postsynaptic target cell, where they bind to protein molecules called *receptors* that are located on the cell surface. Each class of transmitter molecule is capable of being recognized by a number of different receptors, and these fall into two general classes: excitatory and inhibitory. If the target cell has excitatory receptors for a given transmitter, the binding of the transmitter to receptors will increase the probability that an action potential will be initiated in the target cell. Conversely, if the cell has inhibitory receptors, these receptors will actively prevent the initiation of action potentials. The same target cell typically will have excitatory receptors for some transmitters and inhibitory receptors for others.

RAMÓN Y CAJAL PROPOSES THAT SYNAPSES ARE MODIFIABLE

Ramón y Cajal had found that nerve cells are interconnected in surprisingly precise patterns. A given neuron will always connect with certain neurons and not others. We now know that this precision is built into the brain by the precise expression of various genes during development. This precision in neural connections poses an interesting paradox: Presumably nerve cells undergo some sort of change when we learn or remember, but if the connections between neurons are so precisely arranged, what could that change be? How is a precisely wired set of connections modified by neural activity? Do learning and memory require some further additions to the wiring diagram?

In a remarkable insight, Ramón y Cajal proposed a solution to this dilemma. He formulated a hypothesis, now called the *synaptic plasticity hypothesis*, that says that the *strength* of synaptic connections—the ease with which an action potential in one cell excites (or inhibits) its target cell—is not fixed but is plastic and modifiable. Specifically, he postulated that synaptic strength can be modified by neural activity. He further suggested that learning might make use of the malleability of synapses. Learning would produce prolonged changes in the strength of synaptic connections by causing the growth of *new* synaptic processes, and the persistence of these anatomical changes could serve as the mechanism for memory.

Ramón y Cajal spelled out this idea in his Croonian Lecture to the Royal Society in 1894:

> Mental exercise facilitates a greater development of . . . the nervous collaterals in the part of the brain in use. In this way, preexisting connections between groups of cells could be reinforced by multiplication of the terminal branches. . . .

Ramón y Cajal predicted that the learning process might alter the patterns and intensities of electrical signals that constitute the brain's activity. As a result of this altered activity, neurons should be able to modulate their ability to communicate with one another. The persistence of these alterations in basic synaptic communication—a functional property called *synaptic plasticity*—might provide the elementary mechanisms for memory storage. The first test of this idea in a whole animal had to wait until a full 75 years after Ramón y Cajal proposed it, when the study of habituation achieved a breakthrough.

A SIMPLE CASE OF SYNAPTIC PLASTICITY

The first attempt at the neural analysis of habituation was undertaken as early as 1908, in the course of studies focusing on the isolated spinal cord of the cat. The spinal cord controls a multitude of reflex responses that underlie posture and locomotion. For example, a cat withdraws a limb when that limb is touched. The British physiologist Sir Charles Sherrington found that this reflex decreased with repeated stimulation and recovered only after many seconds of rest. Sherrington was greatly influenced by Ramón y Cajal's work, and he tried to develop his notions of reflex action in relation to Ramón y Cajal's anatomical findings. In fact, it was Sherrington who actually coined the term "synapse" (which comes from the Greek word meaning to clasp or to embrace). Sherrington had the great insight to appreciate that a *plastic* change at synapses, along the line suggested by Ramón y Cajal, could be responsible for the habituation he observed in the limb withdrawal reflex. But he could not test this intriguing hypothesis with the neurophysiological techniques then available.

Our understanding of the problem was advanced an important step further in 1966 by Alden Spencer and Richard Thompson, working at the University of Oregon. In a beautiful series of behavioral experiments, they found close parallels between habituation of the limb withdrawal reflex in the isolated spinal cord of the cat and habituation of more complex behavioral responses in intact animals. They thus felt confident

that habituation of spinal reflexes was a good model for studying habituation in general. The limb withdrawal reflex is initiated by activating touch-sensitive sensory neurons that receive information from the skin of the cat's hindlimb. These sensory neurons all send their axons into the spinal cord, where they activate a series of excitatory and inhibitory neurons. Signals from these neurons then converge on the motor neurons, whose activity causes the cat to withdraw its limb. By recording the activity of individual

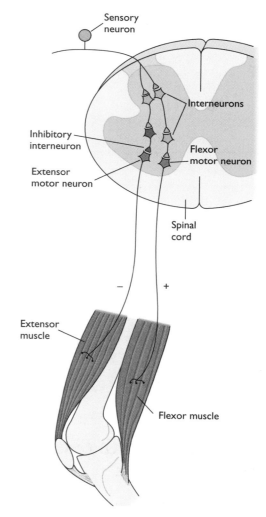

The neural circuit for the limb withdrawal reflex of the cat.

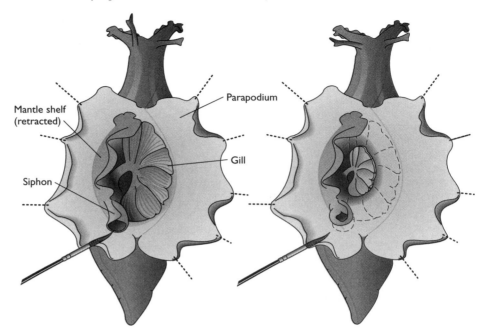

Mantle shelf (retracted)

Parapodium

Gill

Siphon

The gill and siphon withdrawal reflex of Aplysia. *A light touch to the siphon with a fine paintbrush (left) causes the siphon to contract and the gill to withdraw under the protection of the mantle shelf, here shown retracted for a better view.*

motor neurons in the spinal cord of cats, Spencer and Thompson found that habituation leads to a decrease in synaptic activity somewhere in the population of neurons, called *interneurons,* which are interposed between the sensory nerves that detect touch and the motor neurons that signal muscles to contract. The organization of the interneurons in the spinal cord proved quite complex, however, and difficult to examine. Therefore, the synapses critically involved in habituation could not be isolated. These and related studies made it clear that scientists required still simpler systems if they were to analyze habituation or other forms of learning further.

A number of researchers turned to invertebrate animals such as snails and flies because the nervous systems of these animals contain a relatively small number of cells, thereby simplifying the task of a cellular analysis. As we saw in the first chapter, the nervous system of the marine snail *Aplysia* contains only 20,000 cells, many of them are unusually large (some reach almost one

millimeter in diameter). In addition, many of the cells are invariant and identifiable; they can be named and recognized in every member of the species. Thus, the same cell can be studied in both naive animals and animals that have been trained on a particular task.

Eric Kandel and Irving Kupfermann realized that *Aplysia* has a defensive withdrawal reflex that is in some ways analogous to the limb withdrawal reflex in the cat. The creature has an external respiratory organ, the gill, that normally is only partially covered by the mantle shelf, a covering that contains the snail's thin, internal shell. The mantle shelf has a fleshy continuation called the siphon. When either the mantle shelf or the siphon is gently touched, the siphon contracts and the gill withdraws briskly into a cavity underneath the mantle shelf. The defensive purpose of this reflex is clear: it protects the delicate gill from possible damage. Like other defensive responses, the gill withdrawal reflex habituates when it is repeatedly elicited by applying a weak,

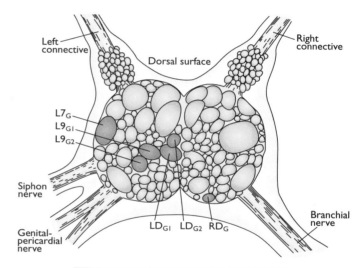

Short-term habituation and spontaneous recovery of the gill withdrawal reflex. A photocell records the movement of a gill retracting in response to siphon stimulation. The movement is then recorded as curves like those shown above. Records from a single long training session, consisting of 79 repetitions of the stimulus at three-minute intervals, show the gill withdrawal response habituating. Most of the decrease in response occurred within the first 10 stimuli. Following a two-hour rest the response recovered partially.

harmless stimulus to the siphon. Typically the experimenter touches the siphon with a fine paintbrush, causing both siphon and gill to withdraw briskly. After one training session of 10 stimuli applied to the siphon, an animal shows very little or no reaction to the tenth stimulus: a touch to the siphon will now cause little or no withdrawal of either the siphon or the gill. The duration of the memory for habituation is a function of repetition. Following 10 stimuli to the siphon, the memory is short-lived—it will last only 10 to 15 minutes. By contrast, after four such training sessions of 10 stimuli each, spaced over four days, the memory for habituation will be prolonged and will last three weeks. The former case is an example of short-term memory for habituation, whereas the latter case is an example of long-term memory.

Biologists noted the similarity between habituation in *Aplysia* and habituation in mammals, including humans. That similarity made it attractive to use *Aplysia* to address three questions: What is the locus of memory storage for habituation in the nervous system? Do plastic changes in elementary synaptic connections contribute to memory storage? If so, what are the cellular mechanisms for memory storage? The answers to these questions would be expected to shed considerable light on simple forms of memory in the animal kingdom. But to find these answers, investigators first needed to work out the wiring diagram of the gill withdrawal reflex.

In invertebrates, the central nervous system consists of collections of nerve cells called ganglia. *Aplysia* has 10 such ganglia in its central nervous system. The gill withdrawal reflex is controlled by one of these ganglia, the abdominal ganglion. This ganglion contains only about 2000 cells, yet it is capable of generating not simply one but a number of different behaviors: siphon withdrawal, respiratory pumping, inking, mucus

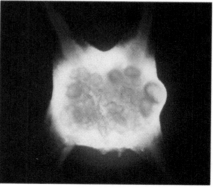

Top: *A map of the dorsal surface of the abdominal ganglion in* Aplysia *shows the location of the six gill motor neurons (in maroon) that participate in the gill withdrawal reflex. The neurons have been labeled L or R, for left or right hemiganglion, and assigned a number. The six labeled cells have a subscript G to indicate their behavioral function as gill motor neurons. Bottom: A photomicrograph of the abdominal ganglion in* Aplysia.

This simplified circuit shows key elements involved in the gill withdrawal circuit. About 40 sensory neurons in the abdominal ganglion innervate the siphon skin. Only eight are illustrated here. These sensory cells terminate on a cluster of six motor neurons that innervate the gill and on several groups of excitatory and inhibitory interneurons that synapse on the motor neurons. (For simplicity, only one of each type of interneuron is illustrated here.)

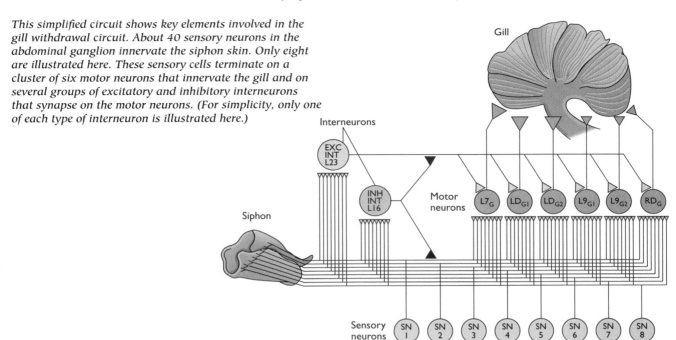

release, egg laying, increases in heart rate, and increases in blood flow. The number of nerve cells that are critical for controlling the gill withdrawal reflex is relatively small—about 100. Thus individual cells make a significant contribution to the overall behavior.

The neural circuit of this behavior was delineated in good part in the early 1970s by Kandel and his colleagues Kupfermann, Vincent Castellucci, Jack Byrne, Tom Carew, and Robert Hawkins, working at Columbia University. In the course of their work, they identified many of the cells of the gill withdrawal circuit. They found six motor cells innervating the gill and seven motor cells innervating the siphon. These motor cells receive information directly (monosynaptically) from two related clusters of about 40 sensory neurons that innervate the siphon skin. The sensory neurons also connect to clusters of excitatory and inhibitory interneurons that in turn project to the motor cells. Thus, stimulating the siphon skin activates the sensory neurons, and

the sensory neurons activate the gill and siphon motor neurons directly. The sensory neurons also activate various interneurons that in turn connect to the motor cells.

The cells of this neural circuit as well as their interconnections are always the same. In all individuals, a given cell will always connect to certain cells and not to others. With their knowledge of the neural circuit, researchers could now address the paradox we considered earlier: How can learning occur and memory be stored in a prewired neural circuit? Kandel and his colleagues were now in a position to resolve this paradox. They found that it has a rather straightforward solution. Although the pattern of connections of the gill withdrawal reflex is set once and for all early in development, the precise strength of the connections is not. We shall here focus only on the gill withdrawal component of the reflex initiated by stimulation by the siphon; but similar rules apply to the withdrawal reflex of the siphon itself when it is touched. In response to a novel

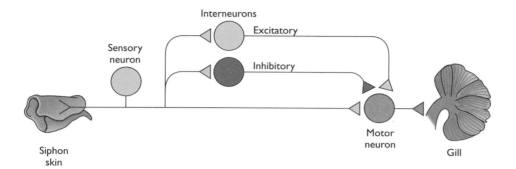

A highly schematized circuit for the gill withdrawal reflex, showing only one of each type of neuron involved.

stimulus to the siphon, the sensory neurons receiving information from the siphon excite the interneurons and gill motor cells quite powerfully. These combined inputs converge on the motor cells to cause them to discharge repeatedly, leading to a brisk reflex withdrawal of the gill. If the stimulus is now repeated, the gill withdrawal reflex response habituates. An action potential in any of the sensory neurons still produces an excitatory synaptic potential in both kinds of target cells—the interneurons and motor neurons. But that synaptic potential is weaker—so weak as to trigger only a very few and finally no action potentials in the target cells. As the connections between the sensory neurons and their targets become weaker with repeated tactile stimuli to the siphon, an action potential in the sensory neurons does not as readily produce an action potential in either the interneurons or the motor neurons. In addition, the excitatory synaptic connections made by some of the excitatory interneurons onto the motor neurons also become weaker. The net result of all this weakening in synaptic connections is that the magnitude of the gill withdrawal reflex response is reduced.

Because this very first relay point of the reflex—the connection between the sensory neurons and their target cells—is modified in habituation, it became possible to use this component of the reflex as a test system to explore in detail what happens during habituation. Castellucci, Kandel, and their colleagues examined the synaptic depression that occurs in the connection between a sensory neuron and a motor cell. They found that during one training session, in which 10 stimuli were applied to the sensory neuron, there was a dramatic weakening of the synaptic connection that persisted for minutes. A second training session produced a further and more persistent weakening. Depending on the number of training sessions, the synaptic depression can last from minutes to hours (and as we shall see below for even longer), but it invariably lasts precisely as long as the behavioral habituation itself. As soon as the synapses regain their strength, the animal begins responding to touch by briskly withdrawing its gill and siphon. These studies confirmed that elementary synaptic connections undergo plastic changes as a result of learning and that these changes are persistent and form the cellular basis for short-term memory storage.

Castellucci and his colleagues could now address the next question: What accounts for this plastic change? Are the receptors in the motor neuron less responsive to each quantum of glutamate, the chemical transmitter that is released by the sensory neurons? Or do the synaptic vesicles release fewer packets of neurotransmitter with each presynaptic action potential? Castellucci, Lise Eliot, and Kandel, and subsequently (and independently) Beth Armitage and Steve Siegelbaum at Columbia, found that the decrease in synaptic potentials resulted entirely from a decrease in the number of packets of transmitter re-

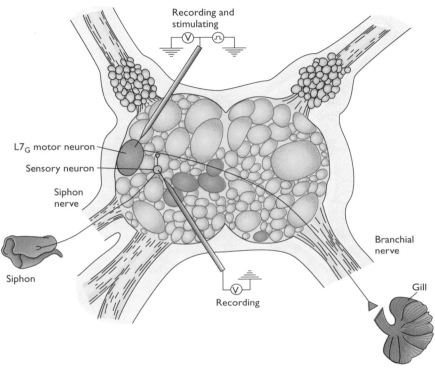

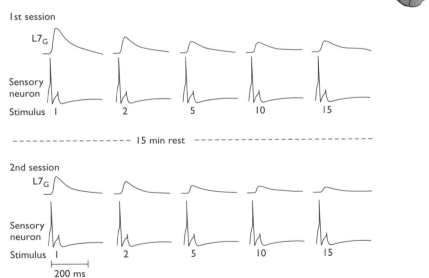

The time course of short-term habituation can be studied by recording the activity of individual sensory and gill motor cells. Top: A sensory neuron that synapses on gill motor neuron $L7_G$ is electrically stimulated every 10 seconds; a microelectrode records the postsynaptic potentials produced in that motor neuron. Bottom: Records from two consecutive training sessions of 15 stimuli, separated by 15 minutes, show that the response in $L7_G$ declines during the first session, is partially restored after a rest, and then declines even more dramatically and almost vanishes with the second training session.

leased by each action potential. There was no change in the sensitivity of the glutamate receptors in the postsynaptic motor neuron.

One feature of the synaptic depression that underlies habituation is particularly interesting: a reduction in transmitter release is already evident in the response to the second stimulus. Whatever molecular events are responsible for the decrease in transmitter release, these events are set in motion as a result of a single stimulus and have already been completed by the time the second stimulus is given. Moreover, the reduction in transmitter release produced by a single stimulus is surprisingly long lasting and persists for 5 to 10 minutes. With the next eight to nine stimuli of a training session, the synaptic depression becomes even more profound and long lasting, persisting for 10 to 15 minutes.

How might the decrease in transmitter release occur? A modeling study by Kevin Gingrich and Jack Byrne at the University of Texas in Houston suggested that the pool of releasable quanta might be depleted as result of habituation. To test this hypothesis directly, Craig Bailey and Mary Chen at Columbia University used the electron microscope to visualize synapses of the *Aplysia* sensory neuron that had been altered by short-term habituation. They found that short-term habituation did not alter the number of presynaptic terminals, the number of active zones in the presynaptic terminals, or the size of the active zones. Nor did it alter the total number of vesicles in a presynaptic terminal. Rather, there was a decrease in the number of synaptic vesicles that were docked at release sites within the active zones, and thus there were fewer packets of transmitter ready to be released. Experiments by Armitage and Siegelbaum suggested that in addition to the reduction in the number of synaptic vesicles docked at release sites, habituation might also interfere with the process by which the remaining vesicles fuse with the membrane of the presynaptic terminal.

These studies illustrate several general principles about memory storage. First, they provided the first direct evidence supporting Ramón y Cajal's prescient suggestion that the synaptic connections between neurons are not fixed but can become modified by learning, and that these modifications in synaptic strength persist and serve as elementary components of memory storage.

Second, we now know what causes the changes in synaptic strength between two critical sets of neurons of the gill withdrawal reflex, the sensory neurons and the motor neurons. In this, the component of the reflex that has been studied most thoroughly, these changes are the result of change in the presynaptic terminals, specifically a change in the number of transmitter vesicles released from those terminals. Although a variety of other plastic mechanisms have been found to contribute to memory storage, altering the amount of transmitter released has proven to be a very common mechanism in creating memory, both in this system and in others.

Third, in the gill withdrawal reflex, there is a decrease in synaptic strength not only in the connections between the sensory neurons and their target cells, but also in the connections made by the interneurons onto their target cells. Thus, memory storage for even a simple nondeclarative memory is distributed through multiple sites.

Finally, the findings illustrate that nondeclarative memory storage does not depend on specialized memory neurons whose only function is to store information. Rather, the capability for simple nondeclarative memory storage is built directly into the synapses connecting the neurons that make up the neural circuit of the behavior being modified. Memory storage results from changes in neurons that are themselves components of the reflex pathway. Thus, the remembrance of habituation is embedded in the neural circuit that produces the behavior. In this respect, as we shall see, nondeclarative memory differs from declarative memory, for which an entire neural system, located in the medial temporal lobe, is designed to help stamp in the remembrance of things past.

THE ADAPTABLE NEURON

We have so far considered short-term memory, the memory that lasts minutes. What about long-term memory, the memory that lasts days, weeks, or longer? One interesting feature of habituation

BEHAVIOR

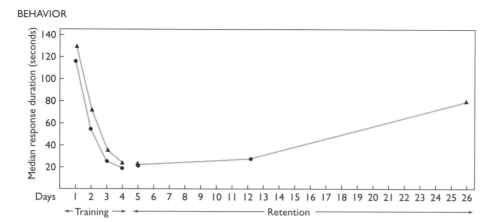

PHYSIOLOGY

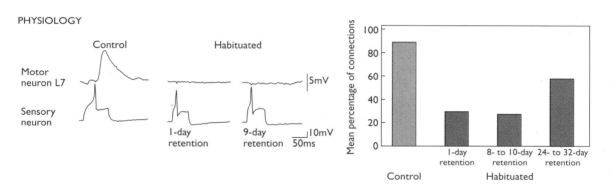

STRUCTURE

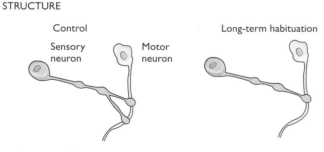

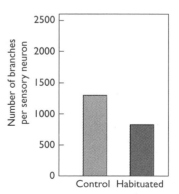

Training sessions of 10 stimuli apiece, given daily over four days to two groups of animals, lead to long-term habituation in Aplysia that persists for more than a week, as shown by the decrease in the duration of the behavioral withdrawal response (top) and by the dramatic depression of synaptic effectiveness observed in recordings of postsynaptic potentials in the motor neurons from these animals (second from top). The time course of the synaptic depression indicated in the top histogram on the right parallels that of the behavioral habituation. This long-term synaptic depression is accompanied by anatomical changes. In the habituated animal the sensory neurons retract their processes and now make fewer contacts onto motor neurons (bottom) than do sensory neurons in control, nonhabituated animals.

of the gill withdrawal reflex is that practice makes perfect. As is the case with other forms of learning, habituation gives rise not only to short-term memory lasting minutes and hours but also, with repetition, to long-term memory lasting days or weeks. Thus, as we mentioned earlier, a single training session of 10 touches to the siphon with a paintbrush gives rise to habituation lasting minutes. By contrast, four training sessions of 10 trials a day, spaced over four days, give rise to habituation that lasts at least three weeks.

One of the key questions in the study of memory is: How do the short- and long-term forms of a memory relate to each other? Do they occur at different loci or at a common locus? To examine this question, Carew, Castellucci, and Kandel gave animals long-term habituation training and then, one day, one week, or three weeks after training, tested the connections between the sensory neurons and motor neurons known to be involved in short-term habituation. They found that in untrained animals about 90 percent of sensory neurons made detectable connections to a given motor neuron. In contrast, in long-term habituated animals only 30 percent of sensory neurons made detectable connections. The remaining connections had been weakened to the point that they could not be clearly detected by electrical recording techniques at one day and one week after training. Paralleling the behavior the connections were partially recovered by three weeks of training.

Here, then, are functioning connections that remain depressed for over one week and are only partially restored by three weeks, and these remarkable changes are the result of a simple learning experience, consisting of four training sessions involving 10 trials each. Thus, whereas short-term habituation involves a transient decrease in synaptic efficacy, long-term habituation produces a more prolonged change—in fact, it inactivates many of the previously existing connections.

How are these profound functional changes maintained? In one of the most surprising and

dramatic findings in the study of long-term memory, Bailey and Chen found that a profound structural change accompanies long-term memory for habituation of the gill withdrawal reflex. The sensory neurons from habituated animals have 35 percent fewer presynaptic terminals than do sensory neurons from control animals. In control animals each sensory neuron has on average a total of 1300 synaptic terminals that contact the total population of target cells—both interneurons and motor neurons. By contrast, the sensory neurons of animals that have been long-term habituated have only about 840 synaptic terminals that contact the population of target cells. This means that in the control condition an average sensory neuron sends about 30 presynaptic terminals to contact each of its target neurons. Following long-term habituation this is reduced to only 20 presynaptic terminals.

These experiments illustrate several additional features of nondeclarative memory. First, the experiments provide direct evidence that, just as short-term memory involves short-term changes in synaptic strength, so long-term memory requires long-term changes in synaptic strength. Second, the same elementary synaptic connections can participate in the storage of both short- and long-term memory. Third, the amount of training necessary to produce a profound change in synaptic function and structure is surprisingly small. Not all synapses in *Aplysia* are plastic and adaptable—some synaptic connections in the nervous system do not change their strength, even with repeated activation. However, at synapses that have evolved to participate in memory storage, a relatively small amount of training—40 appropriately spaced stimuli—can produce large and enduring changes in synaptic strength that persist for weeks.

Finally, the results indicate that synapses are not only plastic in the amount of neurotransmitter they release, but also plastic in their shape and structure. Active zones and presynaptic terminals are not immutable but rather modifiable components of the synapse. The normal set of active zones and transmitter vesicles serves as the anatomical scaffolding for behavior. Even such elementary learning experiences as habituation can alter the scaffolding to modulate the function of neuronal connections. As we shall see in later chapters, these changes in the physical structure of neurons commonly represent an elementary anatomical basis for long-term memory storage.

We have so far considered only the simplest form of nondeclarative memory—the trace in the brain established by learning about the properties of a single stimulus and the decay of this trace as an animal learns to ignore the stimulus. These simple memories are stored as a weakening of the strength of pre-existing synaptic connections. We next turn to slightly more complex examples of learning and ask a further set of questions: Do more complex forms of learning also establish memory traces by altering the strength of synaptic connections? If so, can synaptic connections be strengthened as well as weakened? Finally, we consider the mechanistic underpinning of these storage mechanisms. To understand them deeply, in both health and disease, we need to know: What are the molecular mechanisms whereby synapses are altered in strength?

Jasper Johns, Zero Through Nine *(1961). Johns (1930–) superimposes numbers from zero through nine in order to create an abstract image in which none of the figures is clearly visible. He often uses imagery from everyday life such as targets and flags. Here the superimposition of numbers blurs the distinction between them in much the same manner that layers of stored memories may blur the recall of a specific event.*

3 | Molecules for Short-Term Memory

From the startle reactions of the marine snail *Aplysia,* biologists have learned that a simple form of learning—habituation— leads to a decrease in the strength of synaptic connections and that this decrease, when maintained, serves as a mechanism for storing memory. In this case, the weakening of the synapse results from a single cause: action potentials in the sensory neurons release progressively less transmitter. As a result, the magnitude of the synaptic potential in the target cell—the measure of synaptic strength— decreases with habituation.

These findings, which emerged in the early 1970s, provided the initial evidence for Santiago Ramón y Cajal's suggestion that changes in synaptic strength can contribute to memory storage. In turn, the findings raised a set of new questions, which became the focus of research for the next decade: If habituation produces a weakening of synaptic strength, are there forms of learning that produce an increase in synaptic strength? In revealing that synapses can change, the analysis of habituation had given us a start toward understanding the basis of more complex forms of memory storage but had told us nothing about molecular mechanisms underlying that change. What molecules are important for memory storage? Does learning recruit a novel class of molecules, molecules that are specialized for memory storage, or does memory co-opt molecules used for other purposes?

Now that an elementary type of memory had been localized in synaptic connections, the molecular events underlying memory storage were ripe for analysis. A molecular analysis provides the deepest and most informative insights into the mechanisms whereby all cells, including nerve cells, operate. Moreover, there is the hope that with a molecular perspective we can learn to diagnose and treat diseases that impair the storage of memory, such as Down's syndrome, a disease that affects one infant out of every 100,000, or age-related memory loss, which affects perhaps 25 percent of all people over the age of 65, and the number could well be higher. Since a number of quite different molecular mechanisms could produce similar changes in synaptic strength, to treat a disorder of memory it will be critical to know what mechanisms are involved in normal storage and how specific diseases interfere with normal functioning.

CLUES FROM STUDIES OF SENSITIZATION

The first clues to the molecular mechanisms of a memory process emerged from the study of sensitization, a form of nonassociative learning that results from an *increase* in synaptic strength. With habituation an animal learns about the properties of a *benign* or unimportant stimulus.

With sensitization an animal learns about the properties of a *harmful* or threatening stimulus. An animal that encounters a threatening stimulus learns quickly to respond more vigorously to a variety of *other* stimuli, even harmless ones. A person startled by a gunshot is likely to jump at *any* noise for some minutes thereafter. Similarly, a person will respond more vigorously to a soft pat on the shoulder after receiving a painful shock. By means of sensitization, people and animals learn to sharpen their defensive reflexes in preparation for withdrawal and escape.

In the case of habituation, one has an altered response to a stimulus following repeated presentations of that *same* stimulus. In the case of sensitization, one has an altered response to a stimulus as a consequence of exposure to some *other,* usually noxious stimulus. Thus, sensitization is more complex than habituation and can also override it. For example, a mouse will startle when first exposed to a noise, but as that noise is repeated the mouse will habituate to it and no longer respond. The startle response to the noise can be quickly restored by delivering a single sensitizing shock to the mouse's feet. This ability of sensitization to override a habituated response is called *dishabituation.*

The gill withdrawal reflex of *Aplysia,* so dramatically weakened with habituation, exhibits a robust increase in strength with sensitization. Thus, working at Columbia University, Harold Pinsker, Irving Kupfermann, William Frost, Robert Hawkins, and Kandel found that after an *Aplysia* receives a shock to its tail, its reaction to siphon stimulation is substantially strengthened: it withdraws its gill more completely. It now pulls its gill fully into the mantle cavity, under the protection of the mantle shelf. The animal's memory for this noxious stimulus, as reflected by how long it remembers to enhance its gill withdrawal reflex each time the siphon is touched, becomes more enduring the more often the noxious stimulus is repeated. A single shock to the tail produces a short-term memory that persists for minutes. Four or five shocks produce a long-term memory that lasts for two or more days. Further training produces a memory that lasts for weeks. We will here focus on short-term memory for sensitization, returning to consider its long-term storage in Chapter 7.

If habituation leads to a decrease in synaptic strength, we might wonder if sensitization leads to an increase. And, indeed, Marcello Brunelli, Vincent Castellucci, and Kandel found that applying a noxious stimulus to the tail enhanced a number of synaptic connections within the neural circuit of the gill withdrawal reflex. These included the connections made by the sensory neurons from the siphon skin on both the motor neurons and the interneurons—those neurons that are interposed between sensory and motor cells—as well as the connections made by the interneurons on the motor neurons. This is the same set of synapses depressed by habituation. These studies illustrated that at different times the same set of synaptic connections can be modulated in opposite directions by different forms of learning. As a result, the same set of connections can participate in storing different memories. Synapses that have increased in strength serve as a memory storage site for certain kinds of learning, for example, sensitization, and synapses that have decreased in strength serve as a storage site for other kinds of learning, for example, habituation. In the first case, we say that the synapses have been "facilitated," and in the second they have been "depressed."

Note that, to achieve habituation of the reflex, the mild touch to the siphon activates directly the pathway from the siphon sensory neuron to the gill motor neuron. Thus, the synaptic depression resulting from habituation is *homosynaptic;* it results, as does the behavior, from a change in activity in the same pathway that is excited by the stimulus that elicits the reflex. By contrast, the increase in synaptic strength resulting from sensitization is *heterosynaptic*. The tail shock activates a pathway from the tail that modifies the strength of the connections between the sensory neuron and its target cells, the interneurons and motor neuron. Thus, the increase in synaptic strength is achieved by initially activating a pathway, leading from the tail, that is different from the one that elicits the gill withdrawal reflex, leading from the siphon skin.

A shock to the tail activates sensory neurons located *in the tail*. It does not cause the sensory neurons that innervate the *siphon skin* to fire action potentials, yet somehow it alters the strength of those synapses. How does this occur? Sensitization modifies these synapses through the following steps. A shock to the tail activates sensory neurons in the tail that in turn activate a special class of modulatory interneurons. These interneurons synapse on the sensory neurons carrying information from the siphon skin, making contact on both their cell bodies and presynaptic terminals. Castellucci, Hawkins, and Kandel found that these modulatory interneurons act to regulate transmitter release in the gill withdrawal circuit: they increase the number of glutamate-containing synaptic vesicles that are released each time an action potential is generated in the sensory neurons of the siphon skin. These interneurons are thus termed "modulatory" interneurons because they act to modulate, or tune, the strength of the sensory neuron synapses. As a result of this activity, a mild touch to the siphon that once caused only a small number of vesicles to be released, producing a small synaptic signal, now causes the release of many vesicles, producing a huge synaptic signal in the motor neuron and leading to a more powerful withdrawal of the gill.

There are several types of modulatory interneurons that play a role in sensitization. Each acts similarly, to enhance the release of glutamate-containing synaptic vesicles from a sensory neuron of the siphon, and each does this by engaging the same biochemical signaling machinery *within* the sensory cell.

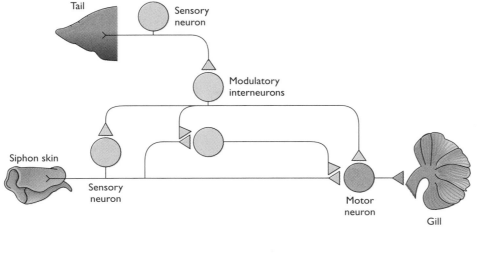

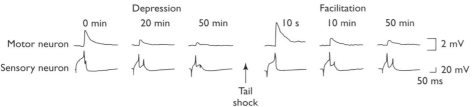

Top: The neural circuit for sensitization of the gill withdrawal reflex in Aplysia *(single neurons are shown for simplicity). A noxious stimulus to the tail activates tail sensory neurons that excite modulating interneurons. Their signals to the sensory neurons of the siphon enhance transmitter release. Bottom: A single synaptic connection can participate in two different forms of memory storage: habituation and sensitization. The synaptic potential produced in the gill by a single siphon sensory neuron undergoes depression as the animal habituates until a shock to the tail sensitizes the animal, and the response is restored.*

The most important of the modulatory interneurons uses serotonin (5-hydroxy-tryptamine, or 5-HT) as its transmitter. Modulatory transmitters such as serotonin that activate the biochemical signaling machinery of their target cells are, as we shall see, critically involved in memory storage not only for relatively simple forms of learning such as sensitization, but also for more complex learning processes. These modulatory transmitters (of which acetylcholine, dopamine, and noradrenaline are other examples) act on receptors on the cell surface of target cells. Learning depends critically on which type of receptor the transmitter binds to and acts upon.

SECOND-MESSENGER SYSTEMS

We have seen that when an action potential triggers the release of neurotransmitter, the synaptic vesicles containing the neurotransmitter fuse with the inside surface of the presynaptic cell membrane in a process called exocytosis. The transmitter molecules then diffuse across the synaptic cleft to interact with receptors in the postsynaptic cell. These postsynaptic receptors fall into two major classes that differ fundamentally in the time course of their action, depending on how the receptor controls the ion channel in the postsynaptic cell.

The mechanism of action of the first class of receptors was discovered in the early 1950s by Bernard Katz and Paul Fatt at University College, London. They found a class of receptors in the postsynaptic cell that is distinguished by the fact that each receptor contains an ion channel directly within its structure. This class of receptors is called *ionotropic receptors,* and the ion channels controlled by these receptors are called *transmitter-gated ion channels.* Ionotropic receptors are responsible for mediating conventional synaptic actions, either excitatory or inhibitory. These synaptic actions are the kind that take place at the synapses in the basic neural circuit that mediates the gill withdrawal reflex. They also take place in other neural circuits that mediate behavior.

This type of synaptic action is fast—generally it persists for only one or at most a few milliseconds. Normally the ion channel of an ionotropic receptor is closed at rest, and ions cannot pass through. When a neurotransmitter such as glutamate is released by the presynaptic neuron, ionotropic receptors recognize and bind molecules of the transmitter. As a result of this binding, each receptor undergoes a conformational change (a change in shape) that opens the ion channel and allows ions to flow into the postsynaptic cell. The flow of ions into the cell produces synaptic potentials, which may either excite or inhibit the cell depending on the type of receptor and ion involved.

In 1959, Earl Sutherland, Theodore Rall, and their students at Western Reserve University in Cleveland and subsequently Paul Greengard at Yale University made the exciting discovery that there exists a second class of receptor. They

IONOTROPIC RECEPTOR   METABOTROPIC RECEPTOR

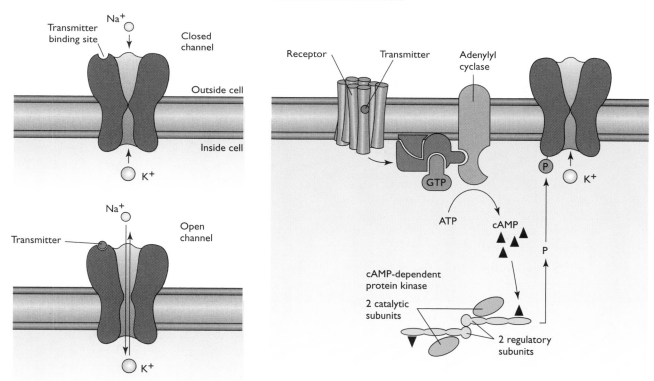

A metabotropic receptor engages a second-messenger system whereas an ionotropic receptor does not. Left: Ionotropic receptors control directly an ion channel that allows sodium (Na+) to enter the cell and potassium (K+) to leave it. By this means, these receptors mediate conventional fast synaptic actions. Right: A metabotropic receptor engages a molecular signaling machinery inside the cell that conveys information from the cell surface to the cell's interior. In this case, by activating the adenylyl cyclase enzyme, the receptor activates a second messenger, cAMP, that in turn activates a protein kinase, the PKA molecule. That kinase then phosphorylates (P) a number of target proteins including an ion channel, causing it to close. This action broadens the action potential, and allows, as we shall see below, more calcium influx and more release of transmitter from the presynaptic terminal.

found that synaptic transmitters can act on a class of receptors that do not contain ion channels. These receptors produce actions in the postsynaptic cell that last much longer than a few milliseconds. They are called *metabotropic receptors,* because they engage the metabolism of the postsynaptic cell—its internal biochemical machinery. As is the case with ionotropic receptors, metabotropic receptors can produce either excitatory or inhibitory actions.

When a transmitter binds to a metabotropic receptor, it activates an enzyme within the cell that alters the concentration of a small intracellular signaling molecule called a *second* or *intracellular messenger.* The function of the second messenger is to convey information into the interior of the cell about the action initiated on the surface of the cell membrane by the transmitter, the *first* or *extracellular messenger.* Second messengers produce their remarkably widespread and prolonged actions because they can affect a variety of functions throughout the cell. However, within a given cell there are a number of second messengers, and each of these is activated by its own set of receptors. These different receptors can be engaged by the same or, more commonly, by different transmitters.

Sutherland and Rall discovered the first-known second messenger: cyclic *a*denosine *mono*phosphate (cAMP). cAMP is related to *a*denosine-*tri*-*p*hosphate (ATP), a ubiquitous molecule essential in all living cells because of the crucial role it plays in almost all biological energy transformations. cAMP is synthesized from ATP by an enzyme called adenylyl cyclase. Metabotropic receptors increase the level of cAMP by activating that enzyme, causing it to convert ATP to cAMP. One of the striking findings about cAMP, which comes from Greengard's work, is that it can affect a variety of biochemical processes within a cell, and it does most of this by activating a single protein—the cAMP-dependent protein kinase (also called protein kinase A or PKA because it was one of the first protein kinases to be discovered).

Kinases are enzymes that add to proteins a phosphate group, a negatively charged chemical group containing phosphorous and oxygen. Adding a phosphate group to a protein—a biochemical reaction called *phosphorylation*—changes the charge and conformation of the protein and thereby alters its activity. Most proteins are activated by phosphorylation; some are inactivated.

How does cAMP activate the cAMP-dependent protein kinase? As is the case with a number of proteins, PKA is a *multimer;* it is made up of several smaller proteins, called *subunits.* In the case of PKA, there are four subunits that are bound together into one multimolecular protein complex. Two of these subunits are catalytic subunits that make up the potentially active component of the kinase enzyme, and two are regulatory subunits that bind to and inhibit the catalytic subunits. As a result, in the resting or basal state of the cell the kinase is inactive. Only the regulatory subunits recognize cAMP. When the concentration of cAMP rises, the regulatory subunits bind cAMP and undergo a shape change, which forces them to release the catalytic subunits. The catalytic subunits are then able to act as an active kinase and phosphorylate target proteins.

Second messengers serve at least three functions. First, they bring the extracellular signal into the cell. Second, they amplify that signal. For example, in a liver cell, where the actions of cAMP were first studied, one molecule of the transmitter epinephrine applied to the outside surface of the cell triggers the release inside the cell of 100 million molecules of the sugar glucose. Third, they regulate not one, but a variety of cellular functions in response to the signal. They produce, as it were, a *state change* in the cell. By means of a second messenger, a small number of synaptic transmitter molecules can trigger a cascading series of biochemical events within the postsynaptic cell. Moreover, other second messengers—such as the ion Ca^{2+}—can modulate the activity of cAMP, enhancing its actions in some cells or counteracting its actions in others.

The cAMP second-messenger system turned out to be crucial in the sensitization of the gill withdrawal reflex. James Schwartz, Howard Cedar, Lise Bernier, and Kandel, as well as Jack Byrne and his colleagues, found that a tail shock stimulates interneurons to release serotonin; serotonin then acts on metabotropic receptors in the sensory neurons to increase the level of cAMP. Even direct application to the sensory neuron of just the modulatory transmitter serotonin on its own increased the level of cAMP. Moreover, the time course of the increase

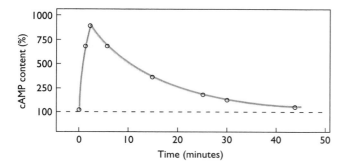

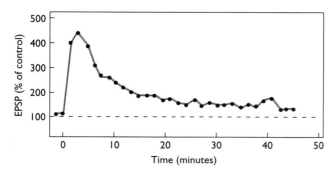

The time course of short-term memory for sensitization parallels the rise in cAMP in the abdominal ganglion of Aplysia. Top: The time course of cAMP increase. Incubating the abdominal ganglion from an Aplysia with a single five-minute pulse of serotonin causes the level of cAMP to increase. Bottom: The time course of the increase in the excitatory postsynaptic potentials (EPSPs) recorded in the motor neurons in response to stimulation of the sensory neuron once every 10 seconds. After a train of stimuli that produces sensitization is applied to a nerve from the tail, the synaptic potential rises and falls with a time course that parallels the amount of cAMP present, suggesting that cAMP plays a role in sensitization.

in cAMP roughly paralleled the time course of short-term memory for sensitization.

Marcello Brunelli, Castellucci, and Kandel next tested whether cAMP was necessary or sufficient to enhance transmitter release. They injected cAMP directly into the sensory neuron and discovered that this injection by itself strengthened the effectiveness of the connection between the sensory neurons and their target cells, producing the same enhancement of transmitter release as an application of serotonin or a tail shock. In collaboration with Paul Greengard, then at Yale University, Castellucci, Schwartz, and Kandel simplified the experiment even further and injected into the sensory neuron a single protein, the catalytic subunit of the PKA. This protein alone enhanced transmitter release. Conversely, injecting into the sensory neurons an agent that inhibited PKA blocked the facilitation. These studies showed that the metabotropic serotonin receptors and the second-messenger system they activated are both necessary and sufficient to strengthen the connections between the sensory and motor neurons in the short term. This cascade is crucial for the synaptic changes that underlie short-term memory for sensitization.

How does the catalytic subunit of PKA act to enhance transmitter release? To address this question, Steven Siegelbaum and Kandel examined some of the target proteins on which cAMP and PKA ordinarily act. They found that serotonin, cAMP, and PKA all act on a novel potassium (K^+) channel, which they called the S channel (because it was modulated by serotonin). This channel is open at rest; it is closed by the action of cAMP. Byrne and his colleagues soon found that serotonin and cAMP also reduce the current flowing through a second class of K^+ channels. Because the K^+ current that flows through these channels is responsible for the duration of the action potential, closing these two types of K^+ channels causes a broadening of the action potential. A broader action potential allows more Ca^{2+} to enter the presynaptic cell, and this Ca^{2+} influx enhances

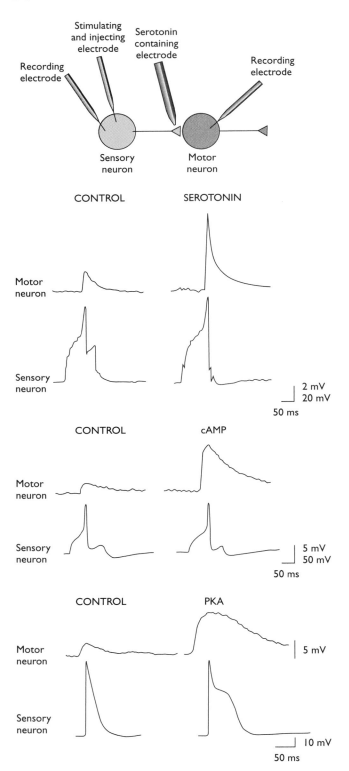

transmitter release. In addition, cAMP and PKA directly enhance transmitter release in a second, Ca^{2+}-independent way, by acting on target proteins that are directly involved in the machinery for vesicle mobilization, fusion, and release.

These studies of sensitization outlined one set of molecular mechanisms that neurons could use to achieve a short-term form of synaptic plasticity. A modulatory transmitter released during learning activates, within critical neurons, a second-messenger signaling pathway that can persist in its activity for a period of minutes. In recruiting cAMP, the second-messenger pathway amplifies the action of the transmitter. Cyclic AMP, acting through PKA, then modulates both ion channels and the machinery for transmitter release. In this way, synaptic connections are strengthened for the full time course of short-term memory. As we shall see later, although different learning processes can recruit different second-messenger systems, the major molecular principles involved in short-term memory are similar in each case.

Taken together, these studies gave new insights clarifying why synapses are such effective and versatile sites for memory storage. Synapses have many faces. They have at their disposal a variety of molecular pathways, able to persist in their activity for varying periods of time, that can either increase or decrease the amount of trans-

In a series of experiments, investigators applied one of three substances to a sensory neuron that synapsed on a gill motor neuron: they applied serotonin to the outside surface of the sensory neuron or they injected cAMP or PKA directly into the sensory neuron. In each case, when they stimulated the sensory neuron with a stimulating electrode, they found that an action potential in the sensory neuron now produced a larger response in the motor neuron. The injection demonstrates another action of PKA: a broadening of the action potential in the sensory neuron as a result of closure of K^+ channels.

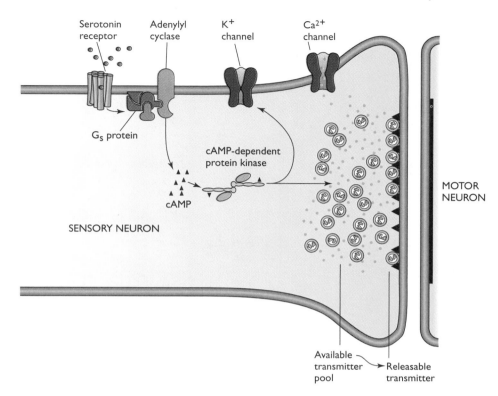

Biochemical steps in the presynaptic facilitation of the sensory neuron. Serotonin binds to a metabotropic receptor, which through a series of steps, involving the coupling protein G_s, leads to the activation of the enzyme adenylyl cyclase, an increase in the level of cAMP, and finally the activation of the cAMP-dependent protein kinase (PKA). The PKA acts on at least two sites. First, it closes down K^+ channels, causing a broadening of the action potential, and a consequent increase in the influx of Ca^{2+} through calcium channels. The broadening of the action potential in turn increases transmitter release. Second, PKA acts directly on some as-yet-unspecified steps in the release machinery.

mitter released by a single action potential. As a result, a single type of synapse is suitable as a storage site for a *variety* of types of memory.

These findings carry with them an interesting philosophical implication. The cAMP pathway is not unique to memory storage. It is not even unique to neurons, as it is used in many other cells of the body—the gut, the kidney, the liver—in order to produce persistent action. In fact, of all known second-messenger systems, the cAMP system is probably the most primitive and is conserved through the course of evolutionary history. It is the only major second-messenger system found in primitive single-celled organisms like bacteria, where it serves as a system to signal

hunger. Thus, memory storage mechanisms in the brain did not evolve through the creation of a specialized set of molecules. Memory does not use a special, memory-related second-messenger system. Rather, memory has co-opted an efficient signaling system that is used for *other* purposes in *other* cells.

In fact, the biochemistry of memory illustrates a rather general biological principle. Evolution does not work by creating new and specialized molecules every time a new and specialized function evolves. Rather, as the molecular biologist Francois Jacob pointed out, evolution is a *tinkerer*. Evolution uses the same collection of genes time and time again in slightly different ways. It does not start from scratch to create new functions as one might in redesigning a computer or a car. Evolution works by creating variations, by generating random changes (mutations) in gene structure, each giving rise to a slightly different variant of a protein. Most mutations are neutral or even detrimental and do not survive. Only rare mutations enhance an individual's survival and reproductive capacity, and these are the mutations most likely to be maintained. Thus, novelty in function is achieved by using pre-existing molecules in a slightly modified form or in novel combinations with other pre-existing proteins. In his book *The Possible and the Actual*, Jacob describes this feature of evolution in the following terms:

> The action of natural selection has often been compared to that of an engineer. This comparison, however, does not seem suitable. First, in contrast to what occurs during evolution, the engineer works according to a preconceived plan. Second, an engineer who prepares a new structure does not necessarily work from older ones. The electric bulb does not derive from the candle nor does the jet engine descend from the internal combustion engine. To produce something new, the engineer has at his disposal original blueprints drawn for that particular occasion, materials and machines specially prepared for that task. Finally, the objects thus produced *de novo* by the engineer, at least by the good engineer, reach the level of perfection made possible by the technology of the time. In contrast, evolution is far from perfection. . . .
>
> . . . evolution does not produce innovations from scratch. It works on what already exists, either transforming a system to give it a new function or combining several systems to produce a more complex one. Natural selection has no analogy with any aspect of human behavior. If one wanted to use a comparison, however, one would have to say that this process resembles not engineering but

tinkering, *bricolage* we say in French. While the engineer's work relies on his having the raw materials and the tools that exactly fit his project, the tinkerer manages with odds and ends. Often without even knowing what he is going to produce, he uses whatever he finds around him, old cardboards, pieces of string, fragments of wood or metal, to make some kind of workable object. . . .

> In some respects, the evolutionary derivation of living organisms resembles this mode of operation. In many instances, and without any well-defined long-term project, the tinkerer picks up an object which happens to be in his stock and gives it an unexpected function. Out of an old car wheel, he will make a fan; from a broken table a parasol. This process is not very different from what evolution performs when it turns a leg into a wing, or a part of a jaw into a piece of ear. This point was already noticed by Darwin . . . . Darwin showed how new structures are elaborated out of pre-existing components, which initially were in charge of achieving a given task but became progressively adapted to different functions. For instance, the glue that originally held pollen to the stigma was slightly modified to affix pollen masses to the body of insects, thus allowing cross-fertilization by insects. Likewise, many structures that make no sense as features subservient to some end and which, according to Darwin, look like "bits of useless anatomy," are readily explained as remnants of some earlier functions. . . .

> . . . Evolution proceeds like a tinkerer who, during millions of years, has slowly modified his products, retouching, cutting, lengthening, using all opportunities to transform and create.

Because the brain is the organ of mental processes, some early students of molecular biology expected to find many new classes of protein molecules in the brain. Instead, there are surprisingly few proteins that are truly unique to the brain, and even fewer signaling pathways—or sequences of communicating proteins—that are truly brain-specific. Almost all proteins in the brain have relatives that serve recognizably similar (homologous) functions in other cells of the body. This is true even for those proteins that we now know to be involved in brain-specific processes—like the protein machinery recruited to release synaptic vesicles or the proteins that serve as ionotropic or metabotropic receptors.

The specificity of the cAMP system is enhanced in at least three ways. First, it gains specificity by the use it makes of the four protein subunits of PKA. We noted earlier that in addition to two catalytic subunits, PKA has two regulatory subunits that inhibit the catalytic subunits of the protein kinase. These regulatory subunits exist in several different forms called isoforms; one of the distinctive functions of the various isoforms is to position the catalytic subunit of PKA in different regions of the cell. Thus, in the sensory neurons, certain isoforms of the regulatory subunits are thought to be localized to the pre-synaptic terminals, and this ensures that the catalytic subunits will also be located there. Second, as a result of this positioning, the catalytic subunit has access to unique target proteins in these cell regions such as K^+ channels and proteins involved in vesicle mobilization and fusion that it cannot access in other cell regions. Finally, because of its location in the presynaptic terminals, PKA also will have the opportunity to interact with other proteins and with other second-messenger systems that operate there. In fact, the cAMP system rarely acts alone. Typically it acts together with another second-messenger system, such as Ca^{2+} or the mitotogen-activated protein kinase system (the MAP kinase), which we shall learn about in Chapter 7, and this cooperation is important for different aspects of memory storage. As a result of these several features, a common cAMP second-messenger system can play a distinctive role in a variety of memory processes.

CLASSICAL CONDITIONING

We have so far considered the two simplest instances of learning: habituation and sensitization. These forms of learning are considered nonassociative because an individual learns about the properties of only a single type of stimulus. To associate two stimuli requires a more complex form of learning called *classical conditioning*. As a rule, classical conditioning can enhance the responsiveness of a reflex more effectively than sensitization, and its effects last longer. How does this occur?

Classical conditioning was first delineated by Ivan Pavlov at the turn of the century. While studying the digestive reflexes of dogs, Pavlov noted that a dog would start to salivate at the sight of an approaching attendant who had fed the dog in the past. The salivation was being triggered by an apparently neutral stimulus, the attendant. Pavlov thus realized that an initially neutral, weak, or otherwise ineffective stimulus could become effective in producing a response as the result of having been associated with a strong stimulus. In this case, the research attendant was the initially ineffective or *conditioned* stimulus (CS), and the attendant was associated with, or paired with, the dog's food, an effective or *unconditioned* stimulus (US). After repeated pairing, Pavlov found that the conditioned stimulus—the attendant—was able to elicit salivation on its own. He therefore called this salivation the *conditioned response* (CR). If the unconditioned stimulus—the food—was now withheld, the conditioned stimulus—the attendant alone—elicited the conditioned response. However, presenting the attendant repeatedly without the food led, after a while, to *extinction,* the gradual decline in the ability of the attendant's presence to elicit salivation.

Pavlov's remarkable discovery was immediately recognized as being extremely fundamental. As early as 350 B.C., the Greek philosopher

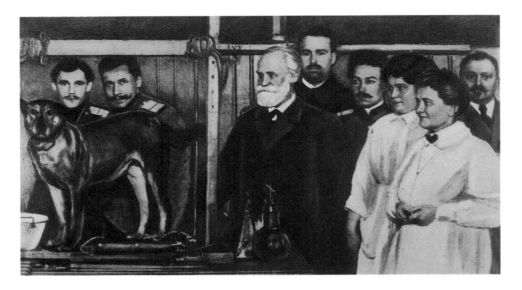

Ivan Pavlov (the white-bearded man at the center) is shown here demonstrating the conditioned reflex in a dog before students of the Russian Military Medical Academy.

Aristotle had suggested that learning involves an association of ideas. This proposal, that we learn by association, was developed further in the eighteenth century by John Locke and the British empiricist philosophers, the forerunners of modern psychologists. Pavlov's brilliant insight was to develop an empirical method for studying the association of ideas by focusing on simple reflex acts and examining the association of two events—two stimuli—instead of the association of ideas.

Since Pavlov's original work, classical conditioning has had a special place in the study of learning. It provides the simplest and clearest example of the rules whereby we learn to associate *two events*. When a person has been conditioned, that person has learned two rules about the events that he or she has learned to associate. The fundamental rule is *temporal contiguity*: the person learns that one event, the conditioned stimulus (CS), precedes by some critical interval a second, reinforcing event, the unconditioned stimulus (US). The second rule is *contingency*: the person learns that the CS predicts the occurrence of the US. This second rule is of particular impor-

tance. Humans and simpler animals need to recognize predictive relationships between events within their environment. They must discriminate food that is edible from food that is poisonous; they must distinguish prey from predator. Humans and other animals can acquire the appropriate knowledge in one of two ways: either the knowledge is innate, and hard wired into the animal's nervous system, or the knowledge is learned through experience. By being able to acquire knowledge through learning, an animal can respond advantageously to a far greater variety of stimuli than it could were it limited to an innate program of knowledge.

The predictive rules that characterize classical conditioning mirror the cause-and-effect rules that govern the external, physical world. It thus seems plausible that animal brains have evolved neuronal mechanisms designed to recognize events that predictively occur together and to distinguish them from events that are not related to each other. It is the existence of these mechanisms in the brain that would explain why animals can be conditioned so readily.

Consistent with this idea is the finding that there is an optimal interval between a particular CS and a US that allows the animal to learn that this CS predicts the US. This optimal interval is surprisingly similar in all species for similar types of associative learning. This constraint suggests that the mechanisms used by neurons to detect contiguity between events may have been conserved through evolutionary history, from snails to flies, to mice, to human beings. For example, in many learning situations involving a noxious unconditioned stimulus, presenting the CS and US simultaneously is not the most effective way to produce conditioning. The best learning results when the CS precedes the US by a short interval and the two stimuli then terminate together. For this type of conditioning, the optimal interval between CS onset and US onset is typically between 200 milliseconds and 1 second. In special cases, the optimal interval may be longer.

What are the neuronal mechanisms that endow animals with the capability to learn and then store in memory the predictive rules of classical conditioning? What accounts for the temporal contiguity of classical conditioning? In exploring the mechanisms of association, we first need to ask: Is the detection of contiguity a property of complex networks, a feature of many cells interacting together? Or is it a feature of individual, specialized cells? If it is a property of individual cells, can it be further reduced to a molecular level? Can molecules important for memory storage have associative properties?

In early studies of learning, it was commonly assumed that associative changes are properties of complex circuits. One of the first people to challenge this assumption was Donald O. Hebb, who boldly suggested that the mechanism for making associations took place inside single cells. In 1949 Hebb proposed that synapses are strengthened by learned associations. This would happen when two interconnected cells are excited simultaneously. Hebb proposed that when activity in the presynaptic cell leads to activity (firing) in the postsynaptic cell, this coincident activity causes a strengthening of the active synapse. In 1965 Kandel and Ladislav Tauc at the Institut Marey in Paris proposed a second cellular mechanism. Synapses become strengthened when the activity of a cell within the CS pathway coincides with the activity of a modulatory neuron that synapses on neurons of the CS pathway. It turns out that in classical conditioning of the gill withdrawal reflex both these mechanisms are used.

THE IMPORTANCE OF TIMING

In 1983 Thomas Carew, Edgar Walters, Robert Hawkins, and Kandel found that the gill withdrawal reflex of *Aplysia* can be classically conditioned. This finding in itself was of considerable interest because it illustrated that even a simple reflex behavior in a relatively simple animal can be altered by associative learning. For classical conditioning in *Aplysia,* a mild touch or very weak electric shock applied to the siphon is used as the CS and a stronger electric current applied to the tail is used as the US. When these two stimuli are paired for about 10 trials, the mild stimulation of the siphon comes to elicit a marked withdrawal of both the gill and the siphon. The withdrawals are significantly larger than if the two stimuli are presented in an unpaired or random fashion during training. This effect builds up during the training session and is retained for several days. As a control, Carew and his colleagues stimulated another CS pathway, this one from the mantle shelf, a skin appendage. The stimulus applied to the mantle shelf was not paired with the tail shock and in fact did not give rise to conditioning.

The classical conditioning occurs if and only if the CS precedes the US by about 0.5 second. It does not occur with an interval of 2, 5, or 10 seconds, nor if the US precedes the CS. This requirement is equally strict for many instances of

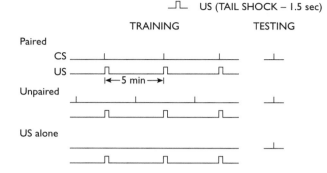

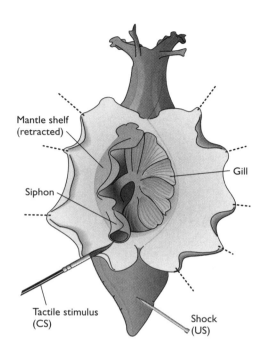

Mantle shelf
(retracted)

Siphon

Gill

Tactile stimulus
(CS)

Shock
(US)

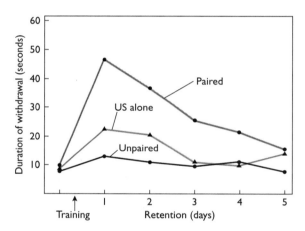

Left: In classical conditioning of Aplysia, *a mild tactile stimulus (the CS) is applied to the siphon and an electrical shock (the US) to the tail. Top right: An experiment compared three groups of animals, receiving three kinds of training: one received the CS and US paired, a second received alternate presentations of the CS and US separated by 2.5 minutes, and a third received only the US. Right bottom: After training, the CS was presented alone once every 24 hours. The group that was trained with both stimuli paired showed the strongest response.*

vertebrate conditioning of defensive reflexes, such as conditioning of the eyeblink response in rabbits, another well-studied case that we shall examine later in Chapter 9.

What happens inside the nervous system to promote the temporal pairing of stimuli? So far biologists understand only the changes in one component of the reflex: the direct connections between the sensory and motor neurons. Here Hawkins, Lise Eliot, Tom Abrams, Carew, and Kandel, and independently Byrne and Walters,

found that the sensory neurons release even more transmitter after conditioning than after sensitization. They called this enhancement of transmitter release *activity-dependent enhancement.* Thus, at least in this component of the reflex, classical conditioning relies on an elaboration of the same mechanism that is used in sensitization.

For behavioral conditioning to occur, the conditioned and unconditioned stimuli must excite the same sensory neurons in sequence and within

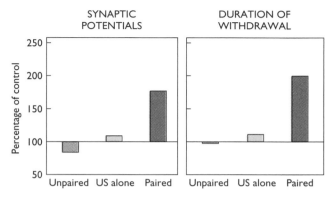

SYNAPTIC POTENTIALS | DURATION OF WITHDRAWAL

A group of animals that was trained with the CS and US paired was the only group that reacted strongly when the CS was subsequently presented alone. The group that had been exposed to both stimuli, in an unpaired fashion, actually habituated to the US and gave a weaker response than normal.

a critical interval. To some extent, classical conditioning proceeds like sensitization: a shock to the tail activates modulatory interneurons that make connections on the terminals of the siphon sensory neurons. A signal from the interneurons, in the form of the transmitter serotonin, enhances transmitter release from the sensory neurons. So far, sensitization and classical conditioning are similar. For classical conditioning to occur, however, the modulatory interneurons cannot just excite the sensory neurons; they must excite them at *just* the right time—just after the sensory neurons

have been excited by the conditioned stimulus in the form of a touch on the skin. This novel property, unique to classical conditioning, is the component that is called *activity dependence*. If, and only if, a mild touch to the siphon excites the siphon sensory neurons *first* and the shock to the tail excites the modulatory interneurons and causes them to act on the sensory neurons *just*

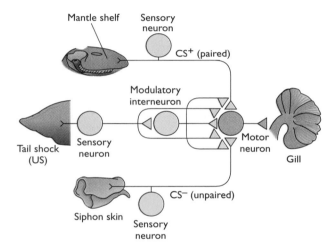

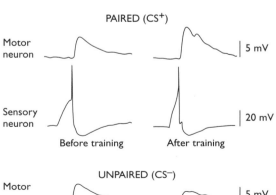

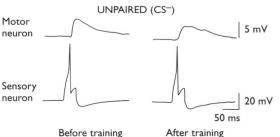

Classical conditioning of the gill withdrawal reflex in Aplysia. *Top: The neural circuit underlying classical conditioning. A shock to the tail excites tail sensory neurons that excite the gill motor neurons directly as well as modulatory interneurons that synapse on the presynaptic terminals of sensory neurons innervating the mantle shelf and siphon. This is the mechanism for sensitization. When a CS is applied to the mantle just before the US, the activity induced in the mantle sensory neurons by the CS makes them more responsive to subsequent stimulation from the interneurons responding to a tail shock. This mechanism contributes importantly to the presynaptic component of classical conditioning. Bottom: Recordings of excitatory postsynaptic potentials, taken before training and an hour later, show that the motor neuron has a stronger response to a signal from the sensory neuron when the CS has been appropriately paired with a US.*

afterward will the sensory neurons show a greater enhancement of transmitter release than occurs with sensitization. If the sensory neurons are thrown into activity by the siphon touch after the tail shock, the tail shock will produce only sensitization. Thus, it is the property of activity dependence that accounts for the precise timing requirements of conditioning.

How do precisely timed stimuli to the tail of an *Aplysia* result in an extra large spurt of transmitter? The answer lies in a series of molecular events that take place between the firing of a sensory neuron and the release of transmitter. There are two components to this process, one presynaptic and the other postsynaptic. We consider the presynaptic component first. As described in Chapter 2, each action potential leads to an influx of Ca^{2+} into the presynaptic terminals. In addition to acting directly on transmitter release, Abrams and Hawkins found that the Ca^{2+} that flows into the presynaptic sensory neuron also binds to a protein called calmodulin. The calcium–calmodulin complex in turn binds to the enzyme adenylyl cyclase, the enzyme that generates cAMP. Once bound to the complex, adenylyl cyclase is more easily activated by the serotonin released in response to a tail shock. As a result, more cAMP is synthesized, more PKA is activated, and more transmitter is released. These experiments showed that the protein molecule adenylyl cyclase has associative properties of the sort predicted by Kandel and Tauc. It will be activated only if signals arrive closely in time. First, the adenylyl cyclase must be primed by the calcium–calmodulin complex; this step is a result of the activity of the sensory neuron. Then the adenylyl cyclase must be activated by serotonin released from an interneuron. As we have seen, serotonin acts on a metabotropic receptor that independently engages the adenylyl cyclase by means of a separate mechanism.

David Glanzman and his colleagues at UCLA and subsequently Jian-Xin Bao, Hawkins, and Kandel at Columbia University, have described a second component of classical conditioning, this one initiated in the postsynaptic cell. One possibility, which we shall consider again in Chapter 6 in the context of declarative memory, is that this change in the postsynaptic cell leads to a signal that travels back to the presynaptic terminals of the sensory neurons, telling them to send more transmitter.

How is the postsynaptic cell changed, and how can that change signal the presynaptic neuron? As we have seen, the sensory neurons use glutamate as a transmitter. The released glutamate, in turn, activates two types of ionotropic receptors: a conventional receptor that is called the *AMPA receptor* (the alpha-amino-3-hydroxy-5-methyl-4 isoxazole proprionic acid receptor) and a special receptor capable of fluxing Ca^{2+}, called the *NMDA receptor* (the N-methyl-D-aspartate receptor). During normal synaptic transmission, as well as during habituation and sensitization, only the conventional AMPA receptor is activated by glutamate because the NMDA receptor channel normally is blocked by magnesium (Mg^{2+}) ions. When the CS and US are appropriately paired, however, the motor neuron generates a whole train of action potentials. These action potentials reduce the electric potential of the motor cell membrane and thereby cause the Mg^{2+} to be expelled from the NMDA receptor channel. As a result, Ca^{2+} rushes through the NMDA receptor channel into the postsynaptic motor neuron. The Ca^{2+} influx into the postsynaptic cell, in turn, activates a set of molecular steps, one consequence of which is thought to be the production of a retrograde messenger that feeds back on the presynaptic cells telling them to boost their release of transmitter even further.

Here, in the NMDA receptor, we see a second molecular associative mechanism, a mechanism of the type predicted by Hebb 50 years ago. The NMDA receptor is activated to allow an influx of Ca^{2+} only if two conditions are met: the receptor must bind glutamate, and it must bind it at a time

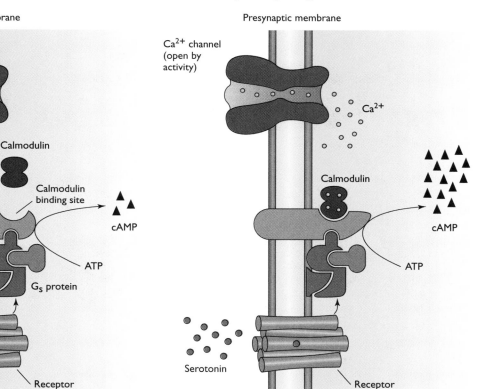

Sensitization

CS⁻ PATHWAY (no preceding activity)

Classical Conditioning

CS⁺ PATHWAY (preceding activity)

The presynaptic component of the molecular mechanisms contributing to classical conditioning. As is indicated in the panel on the right, during classical conditioning, the presynaptic sensory neuron is activated by the CS so that it fires action potentials just before the US. Under these circumstances, the Ca^{2+} influx produced by the action potentials in the sensory neurons forms a complex by Ca^{2+} binding to calmodulin. This complex primes the enzyme adenylyl cyclase. The effect is to make the enzyme more easily activated by the serotonin released by the US. As a result, more cAMP is generated during classical conditioning than during sensitization, in which no preceding activity occurs. When, as is illustrated in the panel on the left, there is no preceding activity in the sensory neurons of the CS pathway at the time the US is activated, adenylyl cyclase is activated less, and less cAMP is generated, which only leads to sensitization.

when the membrane potential is sufficiently reduced to expel Mg^{2+} from the mouth of the channel. When these two conditions are met, as happens when a CS and a US are paired, the Ca^{2+} influx through the NMDA receptor leads to a change in the postsynaptic cell, which is thought to feed a signal back to the presynaptic neuron. Variants of this Hebbian mechanism,

first discovered in the mammalian brain, are also used importantly for storing declarative memories, and we shall consider this mechanism in greater detail in Chapter 6.

These studies of classical conditioning make two crucial points. First, they provide yet another case of the many faces of a single synapse, for we see the same synaptic connections participating in

NMDA CHANNEL CLOSED

Postsynaptic membrane not depolarized

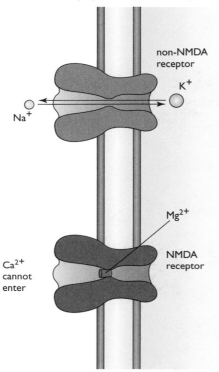

non-NMDA
receptor

K⁺

Na⁺

Mg^{2+}

Ca^{2+}
cannot
enter

NMDA
receptor

NMDA CHANNEL OPEN

Postsynaptic membrane depolarized

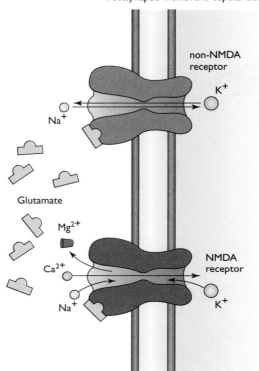

non-NMDA
receptor

K⁺

Na⁺

Glutamate

Mg^{2+}

Ca^{2+}

Na⁺

NMDA
receptor

K⁺

*The postsynaptic component of the molecular mechanism contributing to classical condition-
ing. The left panel indicates the resting condition. As indicated in the right panel, the train of
action potentials produced by the pairing of the CS and US depolarizes the motor neuron
substantially and thereby unplugs the NMDA receptor channel. Calcium flows in, activating
a set of molecular steps, one of which is thought to be the sending of a signal back to the sensory neuron telling it to release even more transmitter.*

still a third variety of learning—habituation, sensitization, and now classical conditioning—and
thereby contributes to still a different memory
storage process. Second, they illustrate that even
rather complex forms of learning and memory
storage use elementary mechanisms of synaptic
plasticity, both pre- and postsynaptic, in combination, almost like a cellular alphabet.

INSIGHTS FROM MEMORY MUTANTS

If the cAMP-based mechanisms for changing
synaptic strength seem intricate, it is because
these mechanisms must have the flexibility to be
used in different ways, so as to serve not only

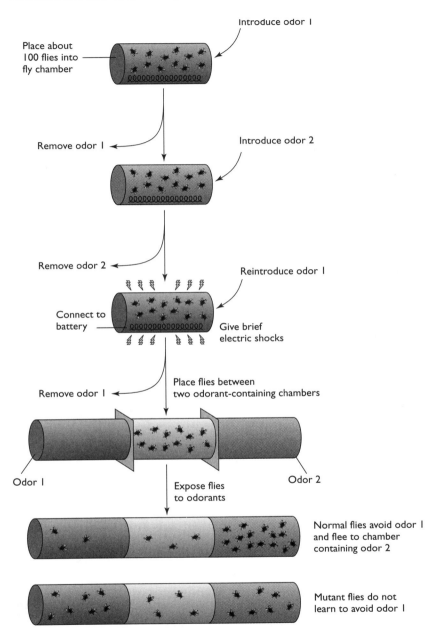

Testing flies for learning and memory. Normal flies remember which odor was paired with a shock and flee that odor. Flies with learning mutations do not flee that odor; instead, they distribute themselves equally among the chambers.

one, but a number of different forms of memory storage. This conviction is based on the remarkable convergence between studies of learning and memory carried out in *Aplysia* and analogous studies carried out in the fruit fly *Drosophila,* using two very different approaches. Whereas studies of *Aplysia* explored the animal's behavior through cell biology, studies of *Drosophila* approached the animal's behavior through the study of genes.

As we saw in Chapter 1, biologists have good reason to work with fruit flies for genetic studies. For one thing, we have a better understanding of the genetics of behavior in the fruit fly than in any other organism. As a result of 90 years of study, it is possible to manipulate the fly genome in a large number of ways: biologists can create mutations in genes, clone mutated genes, and introduce foreign genes. These manipulations have allowed scientists to isolate genes that are critical components of memory storage and that have proven crucial for understanding how memory storage works.

Seymour Benzer, the scientist who originated the genetic study of behavior in *Drosophila,* turned his attention to learning and memory in 1968, when he and his students William Quinn and Yadin Dudai showed that flies are capable of associative classical conditioning. When flies are shocked in the presence of a particular odor, they learn to avoid that odor. Specifically, the flies are placed in a chamber and first exposed to one odor (odor 1) and then to another (odor 2). They are then given an electric shock in the presence of odor 1. Later, the flies are placed in a central chamber with two ends. Each end is a source of one of the two odorants. Normal flies avoid the end containing odor 1, the odor that had been paired with the shock, and stream to the end containing odor 2, the odor that had not been paired with shock. Benzer's students screened thousands of flies to find in-

dividuals that could not remember that odor 1 and shock were paired. In this way they found animals with mutations in genes that affect memory. These memory mutants, rather than avoiding the chamber containing odor 1, will distribute themselves equally among the two chambers. From among these mutants, Duncan Byers, one of Benzer's graduate students, and Ronald Davis isolated *dunce,* the first mutant fly with a defect in short-term memory storage. Remarkably, this fly was found to have a mutation in a gene encoding the enzyme that destroys cAMP. As a result, the animal accumulates too much cAMP, and the synapses become so saturated that they cannot function optimally.

When Quinn, Margaret Livingstone, and Davis went on to search for other mutant flies that had learning deficits, they found that other mutant memory genes also are involved in the cAMP pathway—the same pathway that had been delineated in *Aplysia.* For example, the memory mutant *rutabaga* was found to be deficient in the enzyme adenylyl cyclase, the enzyme that synthesizes cAMP from ATP. The memory mutant *amnesiac* is deficient in a gene for a synaptic peptide transmitter that stimulates adenylyl cyclase, and one mutant called *DCO* has a defect in the catalytic subunit of PKA.

These several findings made it clear that the biochemical mechanism for nondeclarative memory is quite general. It applies to a variety of different forms of learning and to both *Aplysia* and *Drosophila.* This finding encouraged Quinn to focus on PKA as a component—a core signaling pathway—that might be critically involved in a variety of nondeclarative forms of memory. As we have seen, this enzyme alters the activity of various proteins within the cell, including ion channels and the machinery for transmitter release. Quinn was able to express a gene in the fly that shuts down PKA and found that the absence of the kinase interfered with memory storage in the odor task.

In the early studies of learning and memory, Quinn focused on negatively reinforced olfactory discrimination tasks, that is, tasks where the learning was guided by an aversive stimulus. To see whether the cAMP cascade is important in *other* types of learning, Quinn developed a number of learning tasks that required the flies to rely on different senses. For example, he had his flies learn about muscle position instead of odor, react to tasty sucrose instead of electric shock as a reinforcement, or respond by altering posture instead of moving in a particular direction. He then tested normal and mutant flies in these new tasks. In this way, he found that the defects in the mutant flies appeared to be general. Flies that were deficient in one task generally were deficient in all tasks. The simplest interpretation for these findings is that the components of the cAMP pathway that Quinn had affected were essential elements of the biochemical machinery that underlies many types of learning. As a result of the work of Davis, Quinn, and Martin Heisenberg in Germany, about a dozen different forms of learning have now been identified in *Drosophila* that all seem to require the cAMP pathway.

Both cellular studies of *Aplysia* and genetic studies of *Drosophila* indicate that the cAMP cascade is important for certain elementary forms of short-term, nondeclarative memory storage. It is not, however, the only second-messenger system important for synaptic plasticity, as we shall see in Chapter 6. In other instances of learning, and even in variations of sensitization and classical conditioning, other second-messenger cascades play a role.

The surprising and heartening advance that has emerged from the study of invertebrate animals is that one can now pinpoint and observe one type of cellular and molecular machinery, which is used for several different types of learning and short-term memory storage. These studies show that the elementary aspects of various nondeclarative memory processes can be found in the many faces of an individual synapse—in the properties of individual synaptic connections. It will therefore be interesting, both conceptually and technically, to turn next to declarative forms of memory, our memory for facts and events, to see to what degree these more complex forms of memory can be explained in terms of simple synaptic mechanisms. To the extent that such reductionist explanations are possible, it will be interesting to discover how the elementary alphabet of synaptic plasticity may be combined to yield the more complex storage processes that participate in declarative memory for people, places, and objects.

Milton Avery, Girl Writing (1941). A self-taught painter, Avery (1885–1965) developed a style mixing figuration and abstraction. Here the girl at the writing desk uses the faculty of declarative memory, which is essential for virtually all conscious mental activity.

4 | Declarative Memory

Try for a moment to recall the name of a good friend, perhaps a high school classmate or a college roommate. Try to bring to mind the person's face and, if possible, the sound of the voice and the manner of speech. Next, recollect a particular episode that included this friend—an important conversation, a significant event, perhaps a special trip. Re-create the episode in your imagination, moving mentally to the time and place where it occurred. Once the context is reconstructed, it may seem surprising how easy it is to recall the scene and what took place. In this way, one can become immersed in a sustained recollection, sometimes accompanied by strong emotions and by a compelling sense of personal familiarity with what is remembered. Interestingly, in performing an exercise in reminiscence like this, one does not depend on some well-developed retrospective faculty, nor does one need coaching or instruction. Vivid

Declarative memory is a memory of facts.

With its unusual name, this hotel on the Greek island of Santorini seems to invite a mood of nostalgia and remembrance. One expects that here one might form lasting memories or revisit old ones. Our memories are personal and evocative, intertwined with emotion, and they provide us with a sense of who we are.

remembering of the past is something we all do, and we do it every day with minimal effort.

When we speak of bringing a past event to mind, be it a recollection about a friend or the passing thought of some small event from earlier in the day, we are speaking of memory in its most common, most familiar usage. We are speaking of memory as a conscious recollection, as *declarative memory*. In Chapter 1, we introduced the fundamental idea that there are two major forms of memory, declarative and nondeclarative, and that these depend on separate systems of the brain. In Chapters 2 and 3, we considered some simple types of *nondeclarative* memory: habituation, sensitization, and classical conditioning. In this chapter we focus on declarative memory and its several distinct operations: encoding, storage, retrieval, and forgetting.

As we explore declarative memory, we will first be looking, not at cells and molecules, but at aspects of declarative memory that we can observe directly in our own behavior. The ways that we encode, store, retrieve, and forget information provide clues to what declarative memory is and how it functions; and this leads toward an understanding of how declarative memory is organized in the brain.

It is useful to keep in mind that declarative memory does not operate in isolation from other forms of memory. Thus, the same experience can produce many different memories. Consider the case of a single encounter: noticing a dog in the street. You may later recall the scene mentally in a straightforward declarative memory, but you may also experience other aftereffects of this encounter that are revealed later as any of several kinds of nondeclarative memory. For example, on a second encounter, you will identify the animal as a dog more quickly than you did on the first encounter. In addition, depending on what occurred during the encounter, you may develop a phobia or an affection toward dogs, quite independently of how well you remember what happened.

Declarative memory is memory for events, facts, words, faces, music—all the various pieces of knowledge that we have acquired during a lifetime of experience and learning, knowledge that can potentially be *declared*, that is, brought to mind as a verbal proposition or as a mental image. Declarative memory is also called explicit memory or conscious memory. In 1890, the philosopher and psychologist William James described this kind of memory as:

> the knowledge of a former state of mind after it has already once dropped from consciousness; or rather it is the knowledge of an event, or fact, of which we have not been thinking, with the additional consciousness that we have thought or experienced it before.

In a single encounter, we can associate a new name to a new face, learn a story told by a friend, or mentally record the image of a bird we saw perched in the backyard. Sometimes we seem to learn effortlessly and remember for a long time. But such learning and remembering is neither passive nor automatic. Whether or not something that is perceived will be remembered later is determined by a number of factors, the most important of which operate around the time of learning: the number of times the event or fact is repeated, its importance, the extent to which we can organize it and relate it to knowledge that we already have, and the extent to which we rehearse the material after it has first been presented. All these factors influence the nature and extent of the *encoding* that occurs at the time of initial learning and how effectively a new event or a new fact results in neuronal change in the brain. *4 factors for efficient encoding*

ENCODING DECLARATIVE MEMORY

Literally, to encode is to convert information into a code. In psychology the term "encoding" is used in this same sense to refer to the way in which the material we encounter is attended to, processed, and prepared for storage in memory. When encoding is elaborative and deep, memory is much better than when encoding is limited and superficial. One can readily demonstrate this fact by asking two groups of people to study a printed list of eight to twelve simple words. One group is asked to inspect each word and determine the number of letters in the word that are formed entirely by straight lines (e.g., A, E, or H, as opposed to C, R or S). The other group is asked to process the meaning of each word and rate from 1 to 5 how much they like each word. A few minutes later, both groups write down as many of the words as they can remember. The result of the experiment is dramatic and consistent. The group that processed meaning remembers two to three times as many words as the group that focused on the shapes of the letters. The results are similar for other kinds of material such as pictures and musical passages.

In one sense, this finding is trivial. Isn't it obvious that paying attention to the meaning of words should be a more effective way of preparing for a recall test than paying attention to the individual letters? Nevertheless, the experiment illustrates a principle about learning that is pervasive and fundamental. We remember better the more fully we process new subject matter. Memory is better the more we have a reason to study, the more we like what we are studying, and the more we can bring the full breadth of our personality to the moment of learning. Even when learning appears effortless (as in easily being able to recall graduation day or a favorite movie), the learning is not so automatic after all. Particular vignettes and scenes and moments are remembered because they interest us. In these cases, we remember as well as we do because we spontaneously engage in deep and elaborative encoding, as well as rehearsal, that is, we recall the event again and again in our minds.

An example of this processing principle at work appears in Vladimir Nabokov's autobiography *Speak Memory*. Nabokov was best known as a novelist and poet, but he was also an ardent and accomplished student of butterflies who described several new species. His passion for butterflies ensured that certain events would be indelibly recorded.

> And, finally on cold, or even frosty, autumn nights, one could sugar for moths by painting tree trunks with a mixture of molasses, beer, and rum. Through the gusty blackness, one's lantern would illumine the stickily glistening furrows of the bark and two or three large moths upon it imbibing the sweets, their nervous wings half open butterfly fashion . . . *'Catocala adultera!'* I would triumphantly shriek in the direction of the lighted windows of the house as I stumbled home to show my captures to my father.

In contrast, for those less interested in butterflies than Nabokov, not only are such moments not remembered, but the butterflies are not encoded in the first place. He wrote further:

> It is astounding how little the ordinary person notices butterflies. 'None,' calmly replied that sturdy Swiss hiker with Camus in his rucksack when purposely asked by me for the benefit of my incredulous companion if he had seen any butterflies while descending the trail where, a moment before, you and I had been delighting in swarms of them.

When no particular effort is being made to record experiences for later, our interests and preferences direct our attention and determine the quality and quantity of encoding. Our interests and preferences thereby influence the nature and the strength of the resulting memory. In contrast, when we specifically wish to remember, when learning is intentional rather than incidental, we can increase the likelihood of having a strong and long-lasting memory by bringing elaborative encoding processes to the learning task. We can arrange for multiple learning episodes instead of just one, we can rehearse the material to ourselves, and we can build into the learning context retrieval cues that will likely be present when memory is later to be used.

STORING DECLARATIVE MEMORY

Long-term memory is seemingly unlimited in its capacity and can retain many thousands of facts, concepts, and patterns, sometimes for a lifetime. How does the information that is encoded persist as memory? The sequence from perception to memory is best understood in the case of vision, which is the most dominant of the senses in humans and other primates. Indeed, nearly half the cortex is dedicated to processing visual information. More than 30 different brain areas participate, and each area appears to be concerned with particular aspects of the job, for example, the color, shape, motion, orientation, or spatial location of an object.

Whenever an object is perceived, neural activity occurs simultaneously in many different regions. This simultaneous and distributed activity is believed to underlie the visual perception of objects. The question then arises: If the perception of an object is dependent on coordinated activity within widespread areas of cortex, where is memory for the object ultimately stored? The answer is surprisingly straightforward.

There is no separate memory center where memories are permanently stored. Rather, a long line of evidence shows that information storage follows a principle that is conserved across both vertebrates and invertebrates. Memory appears to be stored in the same distributed assembly of brain structures that are engaged in initially perceiving and processing what is to be remembered. There is still no technique that allows one to pinpoint a memory directly in the mammalian brain; we cannot as yet locate the sites where memory of some particular object is stored. Nevertheless, studies of humans and other animals with brain lesions, and newer techniques that allow scientists to obtain images of the human brain at work (functional imaging), have consistently made an important point. The brain regions in cortex that are involved in the perceiving and processing of color, size, shape, and other object attributes are close to, if not identical to, the brain regions important for remembering objects.

Even though no single memory storehouse exists, memory is not spread evenly across the nervous system. It is true that a number of brain regions are involved in representing even a single event, but each region contributes differently to the whole representation. The sum total of changes in the brain that first encoded an experience and that then constitute the record of an experience is called the *engram*.

The principle is that a declarative memory engram is distributed among different brain regions, and these regions are specialized for particular kinds of perception and information processing. This principle helps us understand the extraordinary achievements of persons who are experts in a particular field. Grandmaster chess players can remember chess positions from long-ago matches. Professional athletes can discuss the details of particular plays that occurred within a long match. Tournament Scrabble players can routinely reconstruct the entire board after a match, including the order in which each word was played. However, expert knowledge depends not on the prowess of some general memory talent but on highly specialized abilities, acquired through experience, to encode and organize particular kinds of information. These abilities give experts the ability to recognize quickly a large number of patterns.

In a famous series of studies, William Chase and Herbert Simon at Carnegie Mellon University asked chess players of differing ability to inspect a chessboard displaying authentic game positions and involving up to 26 of the 32 chess pieces. Players viewed the board for five seconds and then attempted to reproduce what they saw on an empty board. Masters and grandmasters could replace about 16 pieces correctly, whereas beginners could replace only about 4 pieces. In the next, critical phase of the study, the chess pieces were arranged randomly on the board so that they did not conform to any real game situation. Under these conditions, the differences between experts and beginners largely disappeared, and all players could replace only 3 or 4 pieces correctly.

Experts have no special gift for remembering details that are not meaningful to their area of expertise. And general memory-training exercises

Chess experts have an exceptional memory for chess board arrangements. Left: The board arrangement after white's twenty-first move in game 10 of the 1985 World Chess Championship in Moscow between A. Karpov (white) and G. Kasparov (black). Right: A random arrangement of the same 28 pieces. After briefly viewing the board from a real game, master players can reconstruct the arrangement of pieces from memory much better than weaker players. With a randomly arranged board, experts and beginners perform the same.

Black

White

Black

White

do not turn a nonexpert into someone who can perform feats of remembering chess pieces. The only relevant training is extended practice within the domain of the expertise itself. The chess expert has stored thousands of chess-piece positions and board configurations and can more easily process those when they recur. Every expert must have accumulated specific changes in the brain that facilitate the perception and analysis of situations relevant to the expertise. The finding that experts have excellent memory for these situations follows naturally from their superb ability to perceive and analyze. Years of practice have changed their brains, and their brains can now encode and process relevant material more fully and in more detail than the brains of nonexperts.

RETRIEVING DECLARATIVE MEMORY

Consider the task of remembering a recently encountered object such as a sports car. Retrieving the object from memory requires bringing together the different kinds of information that are distributed across various cortical sites and reassembling the information into a coherent whole. However, memory retrieval is not simply the reactivation of the various distributed fragments that constitute the engram. Depending on the cue or the reminder that is available, only some fragments of the engram may be activated. If the cue is weak or ambiguous, what is reactivated might even differ from what was stored. For example, some of the reactivated parts could belong to a different episode involving the same sports car or to some different car altogether. One may confuse the thoughts and associations caused directly by the cue with the stored memory content evoked by the cue. The rememberer thus engages in a reconstructive process, not a literal replaying of the past. In the end, a recollective experience may be accepted as accurate and

subjectively compelling when it is only an approximation of the past and not an exact reproduction.

The psychologists Endel Tulving of the University of Toronto and Daniel Schacter of Harvard University have emphasized the importance of retrieval cues. Having a strong memory in storage does not guarantee that you will later retrieve the memory successfully. Suppose you took a weekend family trip with your children to the Grand Canyon and hiked part way to the canyon floor. Years later, some child's casual remark about hiking into a canyon might set off a flood of recollections about your trip. Yet, on another occasion, if your spouse were to ask about the time that a family trip was detoured because roads were closed, and extra hours were spent in traveling, you might not remember what trip this was. Then, as soon as the Grand Canyon is mentioned, the memory of the trip comes flooding back, and perhaps even additional details about the road closures might come to mind. To be effective, retrieval instructions or cues must be able to revivify the memory, and the most effective retrieval cues are ones that awaken the best-encoded aspects of the event that you are trying to remember.

Moods and states of mind can also influence what and how much we remember. The psychologist Gordon Bower of Stanford University found that when student volunteers were given verbal suggestions that induced a sad mood, the students tended to remember negative experiences. In contrast, induction of a happy mood resulted in a bias toward remembering positive experiences. Thus, retrieval is to some extent state-dependent. One's state of mind, indeed the entire context at the time of retrieval, encourages the recollection of events that were earlier encoded in a similar state of mind and in a similar context. Individuals who learn after smoking marijuana or inhaling nitrous oxide (laughing gas) later remember rather poorly what they learned. However, they remember better, though not usually as well as normal, if they are given the same drug again before the retrieval test.

One interesting example of how memory depends on context comes from an experiment on deep-sea divers by Alan Baddeley and Duncan Godden in Cambridge, England. Divers listened to 40 unrelated words, either while standing on the beach or standing under about 10 feet of water. Subsequently, they were tested in one or the other of the two contexts, and asked to recall as many of the words as possible. Words learned underwater were best recalled underwater, and words learned on the beach were best recalled on the beach. Overall the divers recalled about 15 percent more words when the encoding and retrieval contexts were the same.

While these state-dependent effects are intriguing, they should not be overstated, inasmuch as they appear to depend on establishing rather dramatic differences between encoding and retrieval in mood (happy vs. sad), state of mind (drug vs. no drug), or context (on the beach vs. underwater). After all, we need not return to the same room in which we read a book about the Civil War to recall facts from the book. Nevertheless, these findings point out the potential influence of retrieval cues. Overall, retrieval is most successful when the context and the cues that were present when the material was first learned are the same as the context and the cues that are present later when making an attempt to recall. This principle can be usefully applied to ordinary experience. For example, in preparing for an oral examination it is more effective to try to explain the material to another person out loud rather than to simply read the material to oneself.

FORGETTING DECLARATIVE MEMORY

Except in rare cases of isolated and powerful memories like those created on hearing vital news or witnessing an accident, the simple passage of time itself leads to an inevitable weakening of memories that were initially clear and full of detail. With time the details fall away, and we are left with the kernels of the past, the central meanings, not the welter of impressions that were once before us. The day after seeing a film, we might be able to retell the plot and the action in some detail. A year later, it is difficult to recollect more than the gist of the storyline, its mood, and perhaps some fragments of a few scenes.

This loss of memory strength over time refers of course to the familiar phenomenon of ordinary forgetting. On the face of it, forgetting is inconvenient, even a handicap. Wouldn't it be preferable to be able to recall all the material we once studied so hard? Wouldn't it be better never to misplace our eyeglasses or car keys, never to forget where we parked the car, and to hold onto all the events that we view as important? In fact, it is not at all clear that we would be better off if we could remember everything easily. Consider the following story of an unusual individual with a superb, hyperretentive memory.

The Russian neuropsychologist Aleksandr Luria carried out a detailed study of a newspaper reporter named D. C. Shereshevskii, who eventually became a stage memorist. For about 30 years beginning in the mid-1920s, Luria documented the remarkable memory ability of this man, who from early in life had an essentially unlimited memory capacity. Shereshevskii could listen to long lists of words, numbers, and even nonsense syllables and later give back the whole list without error—as long as he was allowed three or four seconds to visualize each item during the learning process. On one occasion he was given a series of letters and numbers presented as a mathematical formula containing about 30 elements. He reproduced the formula correctly on the spot and then reproduced it correctly again 15 years later.

Luria found that Shereshevskii accomplished his remarkable feats of memory by the use of powerful imagery, imagery that came to him involuntarily in response to all sensory impressions.

Words, for example, evoked visual impressions, and sometimes sensations of taste and touch as well. When infrequently Shereshevskii omitted an item from a list, the deficit was in perception or concentration, not in memory. In one instance, for example, Shereshevskii described why he had omitted the words "pencil" and "egg" from a list. His usual technique was to mentally place each image along some familiar street and then to walk along the street, recovering the images.

> I put the image of the *pencil* near a fence. . . . But what happened was that the image fused with that of the fence and I walked right on past. . . . The same thing happened with the word *egg*. I had put it up against a white wall and it blended in. . . . How could I possibly spot a white egg up against a white wall?

> What I do now is to make my images larger. Take the word *egg* . . . now I make it a larger image, and when I lean it up against the wall of a building, I see to it that the place is lit up by having a street lamp nearby.

Despite the advantages of his prodigious recall, there were many serious drawbacks. Shereshevskii's impression of every sensory event was so vivid and so rich that he had difficulty extracting commonalities between events and in grasping the general concepts to form a larger picture. When a story was read to him at a fairly rapid pace, he would struggle against the images that were produced to try to determine the meaning. Words and phrases having different meanings in different contexts (e.g., "case," "blind," "weigh one's words") created particular difficulty. Metaphor and poetry were often beyond him. His memory was so cluttered with detail, and so overwhelmed with separate images, that he could not make use of the kind of organization that focuses on the regularities among several related experiences.

> I frequently have trouble recognizing someone's voice over the phone. It's because a person happens to be someone whose voice changes twenty to thirty times in the course of a day.

The ordinary person's memory is rather different. We are best at generalizing, abstracting, and assembling general knowledge, not at retaining a literal record of particular events. We forget the particulars, and by our forgetfulness gain the possibility of abstracting and retaining the main points. Normal memory is not overwhelmed by the individual and separate details that fill each moment of experience. We can forget the details, and we can therefore form concepts and gradually absorb knowledge by adding up the lessons from different kinds of experiences. It is true that in neurodegenerative conditions like Alzheimer's disease, forgetfulness occurs to a profound and disabling degree. But some degree of forgetting, as it occurs in healthy individuals, is an important and necessary part of normal memory function.

There has been a long debate about what actually happens during forgetting. Do we truly forget or just lose the ability to retrieve memories that still exist in the brain and that might somehow become accessible again? In the late 1960s the psychologists Elizabeth and Geoffrey Loftus at the University of Washington asked 169 people, both laypersons and persons with graduate training in psychology, to choose one of two statements to describe how memory works:

1. Everything that is learned is permanently stored in the mind, and although some particular details are, under normal circumstances, irretrievable, they can eventually be recovered through hypnosis or other special techniques.

2. Some details that are learned may be permanently lost from memory, becoming unrecoverable even with the aid of hypnosis or other special techniques, because the information is simply no longer present.

At that time, 84 percent of the psychologists and 69 percent of the laypersons chose the first statement. In 1996, Stuart Zola and Larry Squire asked the same two questions of 645 health care professionals (nurses, social workers, and clinical psychologists) who were attending workshops on memory. Now 62 percent of them endorsed the first statement. The widespread belief that memory is permanent probably arises from popular notions about hypnosis and psychology as well as the familiar experience that we can often successfully retrieve from memory some apparently forgotten detail from the past. In fact, the popular view is in keeping with Freud's proposal that repression is a major cause of common lapses of memory. Although Freud did consider most forgetting to be psychologically motivated, he also recognized the possibility of literal or biologic forgetting:

> It is always possible that even in the mind some of what is old is effaced or absorbed—whether in the normal course of things or as an exception—to such an extent that it cannot be restored or revivified by any means or that preservation in general is dependent on certain favorable conditions. It is possible, but we know nothing about it.

Of course, opinion polls and debate do not establish the nature of forgetting. Certain biological facts must be determined, for example, whether the cellular and synaptic modifications in the brain that record memory do or do not disappear over time. The information so far available on this point comes mostly from studies of animals with relatively simple nervous systems. The available information suggests that forgetting, whether it occurs after a few hours or many days, is in part an actual loss of information and that some of the synaptic modifications that occurred at the time of learning do, in fact, regress. Contemporary scientists seem to have accepted the idea of literal forgetting. In 1996, Zola and Squire asked 67 scientists with advanced degrees in biology, neuroscience, or experimental psychology to choose between the two statements that appear above. Eighty-seven percent of them chose the second statement, endorsing the idea that some of forgetting involves the actual loss of information.

The most likely scenario is that new information-storage episodes continually resculpt previously existing representations. The effacing of the old by the new, and probably the passage of time itself, change the content of memory. Thus forgetting occurs continuously, weakening and modifying what was learned. However, the gradual disappearance of declarative memory for some event does not mean that no trace of the event is left in the brain. In the first place, some nondeclarative memories may persist, including dispositions and

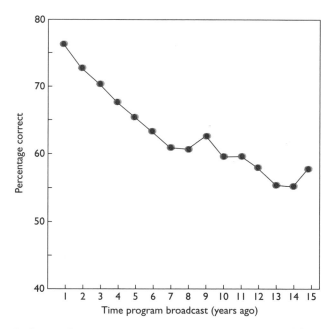

Each year for nine consecutive years (1978–1986), a different group of subjects was tested for their memory of television programs that had broadcast for a single season 1 to 15 years earlier. Although subjects had a better memory for shows broadcast within the last year or two, they nevertheless correctly identified almost 60 percent of the shows from 15 years before. On this test, a score of 25 percent correct could be obtained by guessing.

preferences that were formed as a result of some now-forgotten event, but these are supported by synaptic changes in different regions of the brain from those that supported the declarative memory.

It is also true that when material is relatively well learned initially, forgetting occurs rather gradually and across many years. Even after decades an appreciable amount of memory may still remain. Accurate memories lasting at least several decades and even a lifetime have been documented for important and meaningful information such as the names of one's high school classmates, and even for unimportant information, such as the names of former television programs that had broadcast for only one season.

IMPERFECTIONS IN DECLARATIVE MEMORY

We often do not remember as well as we would like. Memory's frailties are universal to human experience. We may forget an event completely despite our intention to remember it, or we may remember an event inaccurately even when we are sure that we initially perceived it correctly and understood it well. Once some time has passed, our memory of what occurred can grow dim and uncertain. These imperfections in memory can best be understood by considering how memory works and what sort of jobs it is best suited for. Memory does not operate like a tape recorder or video camera, faithfully capturing events for later inspection. Instead, as noted earlier, recollection involves constructing a coherent fabric out of the pieces that are available. When people try to recollect a story, for example, they sometimes make creative errors, deleting some parts of the story, fabricating other parts, and generally trying to reconstruct the information in a way that makes sense. In general, memory

works by extracting the meaning of what we encounter, not by retaining a literal record of it.

In an experiment conducted by John Bransford and Jeffrey Franks, then at the University of Minnesota, subjects studied a set of sentences, including:

1. The ants ate the sweet jelly which was on the table.
2. The rock rolled down the mountain and crushed the tiny hut.
3. The ants in the kitchen ate the jelly.
4. The rock rolled down the mountain and crushed the hut beside the woods.
5. The ants in the kitchen ate the jelly which was on the table.
6. The tiny hut was beside the woods.
7. The jelly was sweet.

Subjects then read several test sentences, deciding in each case whether the same exact sentence had been studied previously. For example, they read:

1. The ants in the kitchen ate the jelly.
2. The ants ate the sweet jelly.
3. The ants ate the jelly beside the woods.

Subjects readily recognized the third test sentence as new. The other two sentences, however, were judged equally familiar, even though only the first one had actually been studied. Subjects apparently abstracted the meaning of the sentences and later could not distinguish between sentences that expressed the same correct ideas about what had been studied.

Once memory is initially established, and the meaning of an event has been recorded more faithfully than the sensory detail, there is further opportunity for the content of memory to change. What is stored in memory can be modified by the acquisition of new, interfering information as well as by later rehearsal and retrieval episodes. Memory may even be modified or distorted by how memory is probed in a retrieval test.

In a study by Elizabeth Loftus and her colleagues, people viewed short films showing automobile collisions. Later some of the viewers were asked: "About how fast were the cars going when they hit each other?" Others were asked the same question but with the key verb changed from "hit" to "smashed," "collided," "bumped," or "contacted." The result was that the average speed estimates given by the viewers were related to how they were queried: smashed (40.8 mph), collided (39.3 mph), bumped (38.1 mph), hit (34.0 mph), and contacted (31.8 mph).

In fact, errors can be introduced into memory at any point: during encoding, during storage, as well as during the act of retrieval. Henry Roediger and Kathleen McDermott at Rice University invited volunteers to listen to a list of words: candy, sour, sugar, tooth, heart, taste, dessert, salt, snack, syrup, eat, and flavor. A few minutes later the subjects wrote down as many words from the list as they could remember. Later, they were also asked to pick out from a longer list of words the ones they had heard earlier and to indicate in each case how confident they were that they had heard it. Forty percent of the subjects wrote down "sweet," even though this word did not appear on the list. Perhaps even more remarkable, when words from the list were presented together with other words, 84 percent of the subjects "recognized" the word "sweet" as one they had heard before, and most of them expressed high confidence that the word was indeed on the list. By comparison, the words that had actually appeared on the list were recognized correctly 86 percent of the time. Thus, the subjects could not discriminate between the words that had been presented and another word ("sweet"), which was related to the list words but had not actually been presented. This experiment demonstrates that it is possible to remember something

that never happened. Presumably, the words on the study list (all words that are closely associated to the word "sweet") evoked a thought of the word "sweet" either at the time of learning or during the memory test, and subjects then confused their merely thinking of the word with actually hearing it. The remarkable conclusion is that it is sometimes difficult to distinguish something that was only imagined from a memory of an actual event.

Children are particularly susceptible to these kinds of influences. In a notable series of studies by Stephen Ceci and his colleagues at Cornell University, preschool children aged three to six participated in several weekly interviews with an adult. Earlier the children's parents had supplied examples of both positive and negative events that had occurred during the past year in their children's lives (e.g., vacation trips, moving to a new house, and injuries requiring stitches). In the interviews, the children were asked to think about some of the real events as well as some other events that, according to the parents, had never happened. Specifically, they were told when each event was brought up, "Try to remember if it really happened." At the end of 10 weeks, the children were interviewed by a new adult. The interviewer brought up in turn each real and fictitious event and asked, "Tell me if this ever happened to you." Depending on the reply, the interviewer then asked for additional details. The main finding was that more than half the children produced false narratives for at least one fictional happening, such as the time they "went on a hot-air balloon ride with their classmates" or the time they got "a finger caught in a mousetrap and had to go to the hospital to get the trap off." Overall, the children agreed that a fictional event had happened about 35 percent of the time that the event was brought up. Their responses were not simple affirmations that certain events had actually happened. Rather, the children told stories, which

In experimental tests, children are questioned about events that may have happened earlier. Children are susceptible to distortions and inaccuracies of memory, especially when subjected to leading questions and false suggestions.

included extensive descriptions of context, and their facial expressions and emotions were appropriate to the story. The following illustrates one of the false narratives from a four-year-old child:

> My brother Colin was trying to get Blowtorch [an action figure] from me and I wouldn't let him take it from me, so he pushed me into the wood pile where the mousetrap was. And then my finger got caught in it. And then we went to the hospital, my mommy, daddy, and Colin drove me there, to the hospital in our van, because it was far away. And the doctor put a bandage on this finger [indicating].

When professionals who work with children were shown videotapes of the children telling their stories, they were unable to distinguish the fictitious stories from the real stories. It seems likely that the stories were so plausible because the children had come to believe that they had actually experienced some of the fictional events. After the experiment was completed, one child responded to his mother's statement that his hand had never been caught in a mousetrap: "But it did happen. I remember it!" If this was typical, then the children were not lying, in the sense of trying to deceive, but were treating events that they had simply thought about as events that had actually happened.

In another study by the same investigators, a stranger introduced as "Sam Stone" visited a preschool classroom for two minutes. He walked around the room, greeted the children, and left. In one condition of the study, the children were given a negative stereotype about Sam Stone before his visit, which suggested that he was very clumsy. In addition, during four separate interviews after his visit, the children were given suggestive questions concerning two fictitious events: "When Sam Stone got the bear dirty, what was that stuff he got on it?" And "When Sam Stone ripped the book, did he do it because he was angry, or by mistake?"

The result was that when the children were eventually questioned by a new interviewer, 72 percent of the three- and four-year-olds claimed that Sam Stone had done one or both of the bad deeds, and 44 percent of them said that they had actually seen him do these things. It is difficult to know whether the children actually believed that the bad deeds had happened, or whether they were saying what they believed the interviewer wanted to hear. In any case, the narratives were elaborate, spontaneous, and full of detail; and again, professionals watching videotapes of the final interview could not determine which children were describing Sam Stone's visit accurately.

Although the distortions and accuracies in memory illustrated here are real features of how memory works, it is also true that memory can be quite accurate. For example, in the case of the Sam Stone study, when children were not given a stereotype, and when the interviews were neutral rather than misleading, most of the children (90 percent) subsequently denied that Sam Stone had done anything to a book or teddy bear. Memory is most accurate when the rememberer is not beset with leading questions or false suggestions, and when the main idea or gist of what happened is being tested rather than the details. Studies of autobiographical recollections of the remote past also document that people usually remember the general meaning and the general texture of past experiences accurately.

Formal experiments show that memory is especially accurate for meaningful visual material. In a famous study, Lionel Standing at Bishop's University in Canada presented volunteers with 10,000 colored, photographic slides that depicted a variety of scenes and subject matter. Each picture was presented only once for five seconds, with rest pauses after every 200 items. Two thousand pictures were presented each day for five days, and at the end of the fifth day memory was tested with a random sample of 160 out of the full set of 10,000 pictures. Each old picture was paired with a new picture, and subjects tried for each pair to choose the picture that they had seen before. Remarkably, subjects averaged 73 percent correct. When one makes a simple correction for the fact that some of the correct answers could have been lucky guesses, one can calculate that subjects were able to remember almost 4600 of the 10,000 slides. It is not known how long all this material could be retained.

This chapter has considered the nature of declarative memory as it is encoded, stored, retrieved, and forgotten. Declarative memory is imperfect, vulnerable to inaccuracy and distortion, but it can also be faithful, especially as an accumulator of general knowledge and as a recorder of general meanings, gist, and main points. The next chapter begins a discussion of how the brain accomplishes declarative memory. Where is memory stored in the short term and in the long term? What brain systems are involved in the encoding, storage, and retrieval of declarative memory, and what jobs do they do?

Louise Nevelson, Black Wall (1960). Nevelson (1900–1984) displays a selection of whole and fragmented found objects in up-ended crates that are fixed together as a unit, reminiscent of the composition of the brain in which separate modules cooperate to support declarative memory.

5 | Brain Systems for Declarative Memory

After having considered declarative memory from a cognitive perspective in Chapter 4, we turn now to the brain systems that support its operation. At this level of analysis, the interplay between cognitive psychology and systems biology is particularly evident. A memory of, say, a charming garden exists as a distributed representation in the cortex, as long as the representation remains in memory—a few moments or many years. In either case, the same cortical areas appear to serve as the storehouse for the representation. But, unlike nondeclarative memory, the transformation of a short-term declarative memory into a long-term declarative memory is not a matter of synaptic connections in just these areas growing stronger. An entirely new brain system comes into play, the medial temporal lobe system. This system is essential for the long-term storage of declarative memory. It is needed at the time of learning, and it remains crucial during a lengthy period of reorganization and stabilization while the ultimate long-term representations are being established in cortex.

SHORT-TERM MEMORY, IMMEDIATE MEMORY, AND WORKING MEMORY

In its most general meaning, the term "short-term memory" refers to the memory processes that retain information only temporarily, until it is either forgotten or becomes incorporated into a more stable, potentially permanent long-term store. Cognitive psychologists subdivide short-term memory into two major components: immediate memory and working memory. Immediate memory refers to what can be held actively in mind beginning the moment that information is received. It is the information that forms the focus of current attention and that occupies the current stream of thought. The capacity of immediate memory is quite limited (it can hold approximately seven items), and unless its contents are rehearsed it ordinarily persists for less than 30 seconds. William James captured the essence of immediate memory (or primary memory, as he called it) when he wrote that this kind of memory was

> never lost, its date was never cut off in consciousness from that of the immediately present moment. In fact, it comes to us as belonging to the rearward portion of the present space of time, and not to the genuine past.

In this chapter we will see that the concept of an immediate memory, as envisioned by William James, is central to understanding how the brain supports declarative memory. Ordinarily, a piece of information will slip from your conscious mind within a few seconds, but

William James (1841–1910), American psychologist.

immediate memory can be extended in time and its contents retained for many minutes if you rehearse actively. This extension of immediate memory is called working memory, a term introduced by Alan Baddeley. An object or fact can be represented initially in immediate memory, its representation can be sustained in working memory, and it can ultimately persist as long-term memory.

This chapter will use the terms "immediate memory" and "working memory" rather than the broader term "short-term memory." In fact, the term "short-term memory" refers not only to capacity-limited, immediate memory and the rehearsal systems of working memory. Short-term memory also refers to even later components of memory, up to the time of the establishment of stable long-term memory. In this sense, short-term memory can persist for many minutes, perhaps an hour or more, well beyond the point when information is actively being held in mind. We describe the cellular and molecular events that lead from short-term memory to stable, long-term memory in Chapter 7.

There is no single temporary memory store through which all information moves on its way to long-term memory. Immediate memory and working memory are best thought of as a collection of temporary memory capacities that operate in parallel. One kind of working memory, the phonological loop, is concerned with language and temporarily stores spoken words and meaningful sounds. This system supports, for example, the ability to hold in mind a telephone number while preparing to dial it and the ability to hold in mind words while speaking or comprehending an ordinary sentence. Another kind of working memory, the visuospatial sketch pad, stores visual images such as faces and spatial layouts. The phonological loop and the visuospatial sketch pad are thought to operate as systems that maintain information for temporary use.

Any particular test of working memory assesses neither the span of consciousness nor the capacity of some all-purpose memory store. For example, a test of how many spoken digits can be held in mind and then repeated back measures the span of just one kind of working memory (the phonological loop). Other information-processing systems have their own working-memory capacities. It is possible that working memory actually consists of a relatively large number of temporary capacities, each a property of one of the brain's specialized information-processing systems.

Biologists have begun to discover something of how the brain organizes its temporary memory functions through neurophysiological studies of monkeys. In some of the earliest work, Joaquin

Fuster and his colleagues at UCLA trained monkeys to remember a color (the sample color) across a delay of about 16 seconds. At the end of the delay two or more colors appeared, and the animal was rewarded with fruit juice for selecting the original, sample color. This task is called delayed matching-to-sample. During the course of the experiment, Fuster recorded the activity of single neurons from area TE, a higher-order visual area in the temporal lobe thought to be important for the perception of visual objects. He found that many neurons in area TE responded to the sample color when it was first presented, consistent with the role of this region in the analysis of visual perception. But it was especially interesting that many neurons continued to respond during the 16-second delay period, as if the persisting neural activity were a neural correlate of the to-be-remembered stimulus. Neurons that exhibit sustained activation while an animal is holding an item of sensory information in temporary memory have also been found in visual cortex, auditory cortex, and sensorimotor cortex during tasks involving visual stimuli, sounds, and active touch, respectively.

In any one of these cortical areas, this sustained neural activity is thought to signal the participation of that area in a larger network. One important region that is active during many such memory tasks is the frontal cortex. Fuster suggested that the frontal lobes are essential for performing tasks that require holding information in mind for impending action, as well as for retrieving information when memory is being reconstructed for the purpose of making some response. Patricia Goldman-Rakic at Yale University had the insight that this holding function of the frontal lobes seemed to involve what cognitive psychologists meant by the term "working memory." She proposed that the frontal lobes hold material in working memory to guide ongoing behavior and cognition.

working memory has different information-processing system that each has it's own span and capacity.

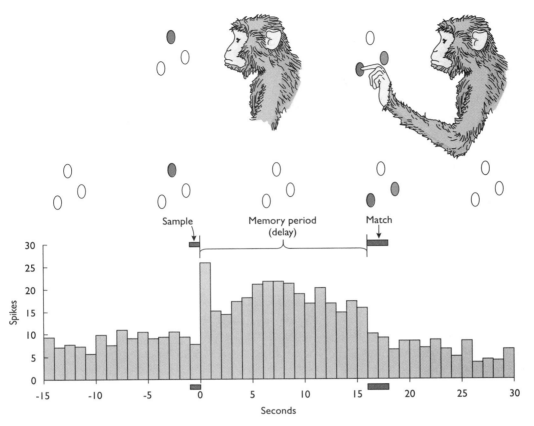

Neuronal activity and working memory. In performing delayed matching-to-sample, a monkey must remember a sample color (here, red) through the subsequent delay ("memory period") to make a proper match and choice of color at the end of the trial. The graph shows the elevated discharge of a cell in cortical area TE during the 16-second memory period. At the end of the memory period, the two color choices (red and green) are presented; the monkey no longer needs to hold the color red in memory, and the activity of the cell drops to the pretrial baseline level.

The frontal cortex has reciprocal anatomical connections with most of the visual areas of the brain, including area TE. Experiments with monkeys show that during the delay period of the delayed matching-to-sample task, which produces neuronal activity in area TE, neurons in one area of the frontal cortex are continuously active. Furthermore, Robert Desimone at the National Institute of Mental Health noted an important difference between the delay activity observed in the frontal cortex and the delay activity in the temporal lobe. He found that he could disrupt delay activity in the temporal lobe by presenting additional visual stimuli during the delay period. Meanwhile, neuronal activity in the frontal cortex continued without interruption. Thus, the delay activity in the frontal cortex may be especially important for holding information in working memory in the face of distraction. Indeed, lesions of the frontal cortex impair a monkey's performance on working-memory tasks. Accordingly, neuronal activity in area TE or other sensory areas is thought to signal the sensory information that is being received at any particular moment. Subsequent "top-down" feedback from the frontal cortex is then thought to sustain neuronal activity in the sensory areas across a delay, biasing these areas toward stimuli that are important for ongoing behavior and that need to be held in working memory. In this way, the frontal cortex and sensory cortices work together as a neuronal system to perceive information and then hold it in working memory for temporary use.

Ultimately, it should be possible to identify a specific set of sensory cortical areas for each distinct working-memory capability that is defined by psychological criteria. Each set of cortical areas would be under the "top-down" influence of the frontal lobes. The frontal cortex would receive sensory information from these upstream cortical areas and then—depending on attention, motivation, and the overall direction of behavior—provide feedback to some subset of these

areas, directing them to hold information in mind for impending action, for comprehension and planning, and possibly for integration into long-term memory.

LONG-TERM MEMORY

Consider the problem of seeing an object and then storing it in long-term memory. The primate visual system is organized in such a way that information from the retina first arrives at the back of the brain in area V1. Visual processing then moves forward from area V1, following two major routes, a "ventral" stream through the lower part of the brain and a "dorsal" stream through the upper part. One route follows a ventral pathway into the temporal lobe, eventually reaching the inferior temporal cortex (area TE), a higher-order visual area concerned especially with analyzing the visual form and quality of objects. A second stream of

Frontal cortex receives info from sensory Areas and based on the attention & behaviour, sends feed back to same sensory area for ongoing behaviour.

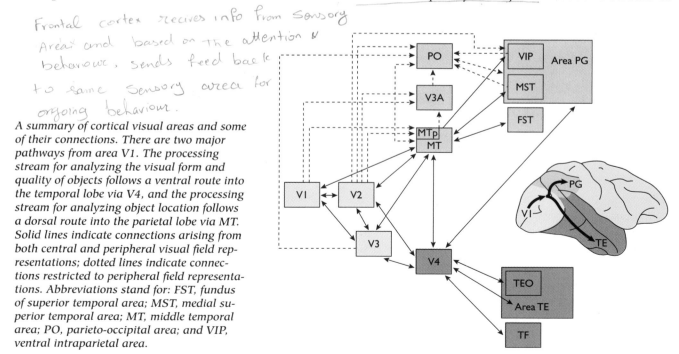

A summary of cortical visual areas and some of their connections. There are two major pathways from area V1. The processing stream for analyzing the visual form and quality of objects follows a ventral route into the temporal lobe via V4, and the processing stream for analyzing object location follows a dorsal route into the parietal lobe via MT. Solid lines indicate connections arising from both central and peripheral visual field representations; dotted lines indicate connections restricted to peripheral field representations. Abbreviations stand for: FST, fundus of superior temporal area; MST, medial superior temporal area; MT, middle temporal area; PO, parieto-occipital area; and VIP, ventral intraparietal area.

visual information processing proceeds by a dorsal route from area V1 forward to the parietal cortex (area PG). Area PG is concerned with the locations of objects in space, the spatial relationship between objects, and the computations needed to reach particular locations in space. Each station along the ventral and dorsal streams is thought to contribute in a specialized way to the processing of information necessary for visual perception. Some areas analyze color, others analyze direction of motion, still others analyze depth or orientation. The more forward areas tend to be more involved with the analysis of whole percepts such as objects. Thus, areas distributed throughout both ventral and dorsal streams are activated simultaneously when we perceive an object in space. What is perceived can persist as working memory when there is sustained neural activity in the same regions, in coordination with activity in the frontal cortex.

This same information can also enter long-term memory, a process that we shall see depends crucially on structures in the medial temporal lobe. Nevertheless, the medial temporal lobe is not the ultimate long-term repository of memory. It is thought that long-term memories are stored in the same distributed set of structures that perceive, process, and analyze what is to be remembered. Thus, one should expect that memory for a recently encountered object would be distributed among area TE in the temporal lobe, area PG in the parietal lobe, and other areas. In each of the relevant areas, persistent changes are thought to occur in the strengths of connections among neurons, and as a result neurons respond differently after learning. It is thought that the aggregate activity in the collection of altered neurons comprises the long-term memory of what was perceived.

An interesting study with monkeys illustrates this idea that the same brain areas appear to be used for long-term memory as are used for visual perception and immediate memory. In this study, Kuniyoshi Sakai and Yasushi Miyashita at the University of Tokyo trained monkeys to learn 12 pairs of colored patterns. Twenty-four patterns were arbitrarily arranged into 12 pairs at the beginning of training, and the monkeys had to learn which pairs of patterns went together, that is, 1 and 1', 2 and 2', . . . 12 and 12'. After the monkeys had learned the pairings, one of the 24 patterns was presented alone as a cue (e.g., 2 or 10'), and a few seconds later the monkeys were able to choose its associate (2' or 10) when it appeared together in a two-choice test with one of the other patterns. The investigators recorded neuronal activity from the inferior temporal cortex while monkeys performed the task. The activity of the neurons indicated that many of them "remembered" the pairings.

The evidence that these neurons might be participating in memory storage came from observing, across many trials, how a neuron responded to each of the 24 patterns when it served as the cue. In both trained and untrained animals, neurons typically "prefer" only one or two of the patterns and respond best when just those particular patterns appear. As the result of training, however, many neurons that once responded to a single member of a stimulus pair now responded to either stimulus. In other words, after training, a neuron that tended to respond only to pattern 2 also responded to the other member of that same pair (pattern 2'). Thus, some neurons acquired new, stable response properties that reflected the storage of the 12 associations in long-term memory. As a group, these neurons have changed as the result of training. Thereafter, they carry the long-term memory of which patterns are paired with each other.

Studies of brain-injured patients have illuminated where such long-term memories are located. These studies have revealed a surprising degree of specialization within the cerebral cor-

Neuronal activity and long-term memory. Top: Twelve pairs of visual patterns (1 and 1', 2 and 2', and so forth) used in a pair-association task to test long-term memory in monkeys. The monkeys learn to choose the one pattern that is associated with each cue pattern. Bottom: Each of the two panels shows how a different neuron responded to all 24 patterns. The light bars show the neuron's firing rate in response to the 12 cue patterns (1 through 12), and the dark bars show its firing rate in response to the corresponding patterns (1' through 12') that are associated through training. In the left panel, the neuron fires to both patterns 12 and 12'. In the right panel, the neuron fires to both patterns 5 and 5'. Because the pairings were arbitrary, these results show that neurons could "learn" the pairings.

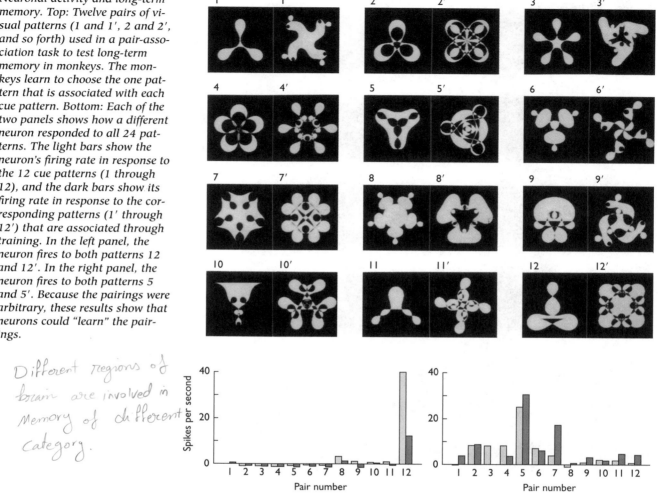

[handwritten margin note: Different regions of brain are involved in Memory of different Category.]

tex. They support the idea that different regions are involved in storing different kinds of memory. Elizabeth Warrington and Rosaleen McCarthy of Queens Square Hospital in London, England, first observed that damage to the left temporo-parietal region of the human brain or to the left fronto-parietal region can produce remarkably selective losses of *category-specific knowledge*. For example, a patient might lose knowledge about one category—small, inanimate objects (brooms, spoons, chairs), but retain knowledge about another category—living things and other large objects (puppies, cars, clouds). Damage to the ventral and anterior temporal lobes, sparing the parietal cortex, can produce the opposite pattern.

These findings led Warrington and McCarthy to suggest that the particular sensory and motor systems that are used to learn about the world influence where in the brain information is ultimately stored. Their idea explains the surprising findings they obtained from their brain-damaged

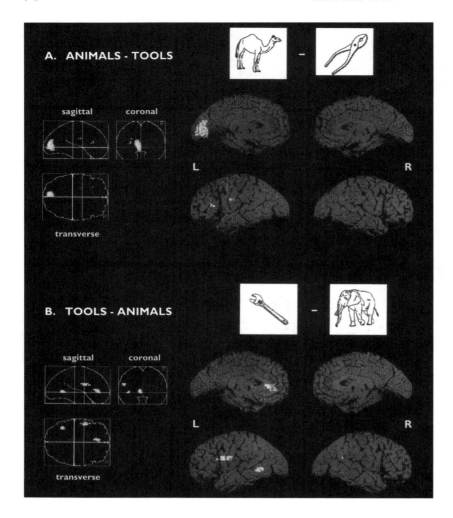

Brain images during naming of animals and tools. These images from PET scans show regions in which cerebral blood flow is increased when subjects silently name drawings of animals compared to when they silently name tools (top) and when subjects silently name tools compared to when they silently name animals (bottom). The images at the left show where blood flow increases occurred when the brain was viewed from the side, front, or top (sagittal, coronal, and transverse, respectively).

proporties of The objects together with how They are percieved influence The representation of long-term memory.

patients. For example, people learn about living things and large, outdoor objects primarily through vision, and many of the brain systems that process shape, color, and visual recognition are located within the temporal lobe. By contrast, people learn about inanimate objects like tools and furniture through processing systems that concern manual interaction and an understanding of function, and the processing systems that relate to manual interaction and knowledge about function are located within the parietal lobe and in the frontal cortex.

Alex Martin, Leslie Ungerleider, and their colleagues at the National Institute of Mental Health used the imaging technique positron emission tomography (PET) to study where category-specific knowledge is stored in the brains of normal subjects. They found that specific cortical areas in the ventral stream of processing were more active during the naming of animals than during the naming of tools, and that other cortical areas, in the dorsal stream, were more active during the naming of tools than during the naming of animals. These results, which agree with the findings of Warrington and McCarthy from patients with lesions, show directly that the properties of objects, together with how they are perceived and used, influence which brain areas store long-term representations about the identity of objects.

MOVING FROM IMMEDIATE MEMORY TO LONG-TERM MEMORY

To consider how long-term memory is established, we return again to the figure on page 87 and the visual processing pathways of the primate brain. The processing pathways in the visual cortex converge on a number of targets, including the cortex of the frontal lobe and the medial surface of the temporal lobe. If any single one of the visual processing areas is damaged, the result is a specific impairment in perception. For example, one lesion may cause a difficulty in the perception of motion and another a difficulty in the perception of faces. An important lesson about the neural organization of memory functions comes from the finding that, in contrast to the effects of a damaged visual processing area (for example, in the lateral temporal lobe), a damaged medial temporal lobe does not impair perception at all. But it has another impact that is far more global: damage to the medial temporal lobe impairs all of declarative memory. As we have seen, memory is a normal consequence of perception. The medial temporal lobe allows for the lasting effects of perceptual experience that we call memory.

To transform a visual perception and immediate memory into a persistent, long-term declarative memory, the medial temporal lobes of the brain must first store aspects of the developing memory. It must then interact with the cortical areas that support perception and immediate memory. One way to appreciate the job performed by the medial temporal lobes is to consider exactly what happens to memory when this region of the brain is damaged. The key observation is that bilateral damage to the medial temporal lobes produces a severe and selective impairment in declarative memory, a clinical syndrome known as amnesia.

AMNESIA

Any condition that damages the medial temporal lobe (or anatomically related areas) may produce amnesia. The cognitive deficit is similar following surgical removal, head injury, stroke, ischemia, anoxia, or disease. The reason that Alzheimer's disease typically begins with symptoms of memory impairment is that the degenerative brain changes characteristic of the disease first appear in the medial temporal lobe. Chronic alcoholism can also result in an amnesic condition, because years of alcohol abuse damage the medial thalamus and the hypothalamus, areas with anatomical connections to the medial temporal lobe.

As we discussed in Chapter 1, the initial insight about the importance of the medial temporal lobes for memory came from observations of the amnesic patient H.M. Since H.M.'s case was first described in 1957, the same clinical picture has been encountered again and again in other patients with bilateral damage to the medial temporal lobe. The hallmark of the memory deficit is profound forgetfulness. It does not matter whether the information to be learned concerns names, places, faces, stories, drawings, odors, objects, or musical passages. And it does not matter whether the material to be learned is presented orally, is read by the patient, or is explored by touch or smell. In all these instances, the patient perceives the material normally and holds it satisfactorily in immediate memory. However, the material cannot persist as long-term memory. Although the principal defect is in the acquisition of new memories, we will see later in this chapter that already-established memories can also be affected.

The material is available so long as it remains in view or is otherwise available to the senses so that it can be perceived. Further, it remains available for use so long as it is being rehearsed and held in working memory. However, once the amnesic patient's attention turns to some other object, then the material is lost from memory. It cannot be retrieved and brought back into view. It has been forgotten.

Damage to the medial temporal lobe spares immediate memory (and working memory), because these early-stage forms of memory are thought to depend on areas of cortex outside the medial temporal lobe. Because of this arrangement, the medial temporal lobe's essential role is not revealed until at least a few seconds have passed after information is presented. After a few seconds, immediate memory (or working memory) can no longer support recollection unless the information is maintained by rehearsal. When recollection begins to depend on later components of memory, the medial temporal lobe becomes essential for memory storage and retrieval.

When the effects of medial temporal lobe lesions were first described in the 1950s, many scientists were skeptical that memory could really be affected in isolation from other cognitive functions. It was more natural to suppose that an impaired memory was the result of some other cognitive problem, such as a clinical depression, inattention, or even a global deficiency of intellectual function. However, Brenda Milner's studies and later experimental work eventually made a convincing case that memory was a separable and isolatable function of the brain. Tests showed that memory-impaired patients could perform normally virtually all functions that did not require new learning. For example, these patients perform well on tests that make large demands on their perceptual abilities. In one such study,

Stephan Hamann and Squire at the University of California, San Diego, measured the minimum exposure time needed to detect words on a computer screen. Common English words were flashed for a duration as short as 33 milliseconds, and subjects were asked to try to identify the words. As the exposure times were gradually increased, control subjects improved from identifying essentially 0 percent of the words correctly (with a 33-millisecond exposure time) to identifying about 90 percent of the words correctly (with a 116-millisecond exposure time). Amnesic patients matched the performance of the control subjects at each exposure time.

Another test, given by Carolyn Cave and Squire, provided an estimate of immediate memory capacity. Subjects, both amnesic and normal, were read a sequence of digits (for example, 5-7-4-1) and then asked immediately to repeat back the sequence. Each time the subject was successful, the number of digits in the test sequence was increased by one (for example, 6-8-2-4-7). Cave and Squire recorded the number of digits, called the "digit span," that was successfully repeated back before a subject failed twice at the same sequence length. Repeated testing resulted in a measure that was accurate to nearly one decimal point. The amnesic patients and the control subjects both repeated back an average of 6.8 digits.

In sharp contrast to the intact perceptual abilities and intact immediate memory of amnesic patients, the deficit in long-term memory can be severe. The problem is best understood as a difficulty in retaining the events of each passing hour and day. Socializing is limited by the difficulty of keeping a topic of conversation in mind and of carrying over the substance of social contacts from one visit to another. Any complex activity presents a challenge because following the correct sequence of steps places a burden on memory. The memory impairment can be documented in the laboratory with any conventional memory test that asks subjects to recall or recognize recently encountered facts or events. The test

Magnetic resonance imaging has revealed the extent of the area surgically removed in the well-known amnesic patient H.M. The diagram at the top is a view of the human brain from below showing the longitudinal extent of the area removed. A through C are drawings of sections through the brain, arranged from front to back, showing the extent of the lesion. Note that although the lesion was made on both sides, the right side is shown intact to illustrate the structures that were removed.

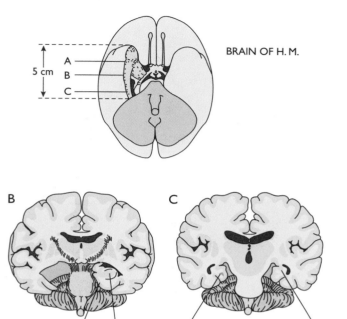

BRAIN OF H. M.

5 cm

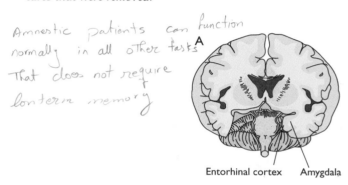

Amnestic patients can function normally in all other tasks That does not require lonterm memory

Entorhinal cortex   Amygdala

Entorhinal cortex   Hippocampus

Small lesion   Hippocampus

need not be framed explicitly as a memory test by asking subjects directly to learn material and then later asking them what they remember from an earlier session. An impairment is also apparent when subjects are simply presented with factual material (e.g., "The Angel Falls are located in Venezuela") and then later asked factual questions without reference to any earlier learning experience (e.g., subjects are simply asked, "Where are the Angel Falls located?").

It is worth distinguishing this kind of amnesia, which results from neurological injury or disease, from functional (or psychogenic) amnesia. Functional amnesia is often described as a loss of personal identity. This kind of amnesia has been popularized in literature and film (for example, the Hitchcock film *Spellbound*), but it is much rarer than the amnesia that results from brain damage and easy to distinguish from it. Functional amnesias typically do not impair new learning capacity. Patients are able to store in memory a continuing record of ongoing events

from the moment they are first encountered by the clinician. The principal symptom of functional amnesia is loss of memory for the past, though patients vary greatly in how this symptom presents itself. Some patients lose personal, autobiographical memory but retain memory for past news events and other facts about the world. Other patients lose personal memory and also lose information about place names, famous people, and facts. For some patients, much of past life is lost; for others, only particular time periods are absent. Emotional factors determine what is lost. Sometimes the functional amnesia passes, and the lost memories are recovered.

THE ANATOMY OF AMNESIA

The medial temporal lobe is a large region of the brain; it includes the amygdala, the hippocampus, and surrounding cortex. H.M.'s lesion involved this entire region, and it was not known at the

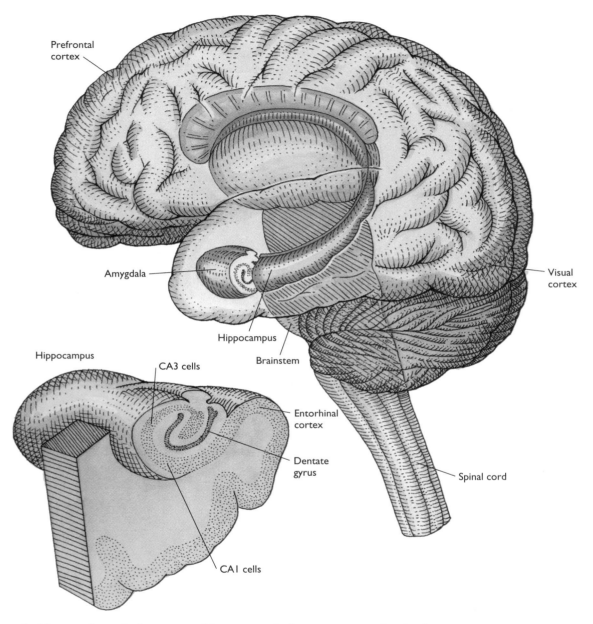

The blow-up shows the hippocampal formation, which is important in forming long-term declarative memories. The CA1 and CA3 cells are part of the hippocampus proper.

This slice through R.B.'s brain shows the hippocampus situated within the medial (or inner) aspect of the temporal lobe. The hippocampus (arrow on right) has a relatively normal appearance except for a region of thinning (between the smaller arrows) in the CA1 region. In this region, virtually all the neurons had been lost.

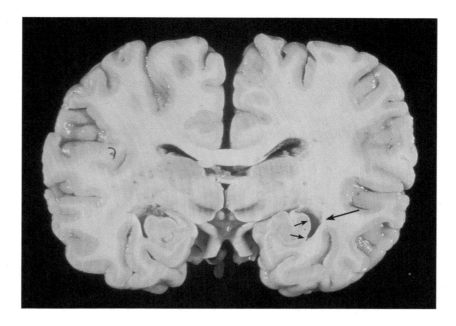

time whether all this region had to be damaged to cause an impairment as severe as his, or whether damage to some relatively small area could be responsible. A few other patients have been successfully studied since H.M. was first described. Their number is small because it is unusual to be able to obtain detailed information about an individual's memory functions over a period of years and then also to obtain a detailed description of the damage from post-mortem examination. Nevertheless, scientists have established some important points about the anatomy of memory from their studies of these patients.

Two patients (G.D. and R.B.) studied by Stuart Zola, Squire, David Amaral, and their colleagues had amnesia caused by an episode of ischemia (a temporary loss of blood supply to the brain as often occurs in a heart attack), and these patients proved to have bilateral damage limited to the CA1 region of the hippocampus. From what is known about the wiring diagram of the hippocampus, a CA1 lesion would be expected to interrupt any contribution that the hippocampus

itself ordinarily makes to memory. These two patients provide the best evidence available that damage limited to even a small region of the hippocampus is sufficient to cause an easily detectable and clinically significant memory impairment. Because the memory impairment in these two cases was only moderately severe, much less severe than in patient H.M., these cases also indicated that other areas of the medial temporal lobe region, beyond the CA1 region of the hippocampus, must be important for memory as well.

Two other patients (L.M. and W.H.) studied by the same investigators, together with Nancy Rempel-Clower, proved to have more extensive damage to the hippocampal formation than G.D. and R.B. The damage in these two patients included all the zones of the hippocampus, including the CA1 zone, as well as a closely related area called the dentate gyrus and an area at the outer edge of the hippocampus called the subiculum (for W.H.). Some cell loss occurred as well in a neighboring area called the entorhinal cortex. Memory in these two patients was correspond-

CHARACTERISTICS OF HUMAN AMNESIA THAT HAVE BEEN PRODUCED IN MONKEYS

1. Memory is impaired on several tasks including ones identical to those failed by amnesic patients.

2. Memory impairment is exacerbated by increasing the retention delay or the amount of material to be learned.

3. Memory impairment is exacerbated by distraction.

4. Memory impairment is not limited to information perceived through only one of the senses.

5. Memory impairment can be enduring.

6. Memory for events prior to the onset of amnesia may be impaired (retrograde amnesia).

7. Skill-based memory is spared.

8. Immediate memory is spared.

ingly more severely impaired than in G.D. and R.B., but still less severely impaired than in H.M. Together, the available cases suggest that memory impairment becomes more severe as more damage occurs within the medial temporal lobe region. However, the limited evidence available from patients has not told us which structures are more important than others, nor what specific damage is responsible for the severe impairment observed in patient H.M. Indeed, soon after H.M. was first described, scientists recognized that it would be essential to develop a model of human amnesia in an experimental animal in order to identify with certainty the structures and connections within the medial temporal lobe that are important for memory. Only with experimental animals can one carry out systematic studies that explore the effects of specific, anatomically restricted lesions on memory and cognition.

AN ANIMAL MODEL OF HUMAN AMNESIA

In the late 1970s, Mortimer Mishkin at the National Institute of Mental Health achieved the first success with an animal model of human memory impairment using the monkey. Initially, he prepared monkeys with large bilateral lesions of the medial temporal lobe to approximate the damage sustained by the amnesic patient H.M. These large lesions eventually proved to reproduce in monkeys many important features of impaired declarative memory in humans, listed in the table on this page. With the animal model in place, it took about 10 years to identify the structures in the medial temporal lobe that are essential for declarative memory. This information confirmed and greatly extended what had been learned from human patients.

In humans, declarative memory is always expressed as conscious recollection. Yet the concept of conscious recollection cannot be studied in monkeys. How can one study in monkeys the kind of memory that is analogous to declarative memory in humans? The answer is that declarative memory has a number of properties other than conscious recollection, and many of these other properties can be studied. Indeed, a memory system that is distinct from others should be separable on the basis of several criteria. For example, it should be possible to distinguish declarative from nondeclarative memory by its operating characteristics, the kind of information that is processed, and the purpose served by the system.

Many memory tasks have been used to study declarative memory in the monkey, but here we

A monkey performs the delayed nonmatching-to-sample task, a test of recognition memory. Left: The monkey is presented with the sample, the red-and-yellow object. Right: After a delay that can span many minutes, the monkey is presented with the sample and a new object. As the test requires, the monkey selects the new object and thereby demonstrates that it has recognized the sample.

will describe only two for purposes of illustration. Human amnesic patients fail these same two tasks when the tasks are given to the patients in exactly the same way that they are given to monkeys. The first task is "delayed nonmatching-to-sample," a simple test of the ability to recognize a recently encountered object as familiar. The task begins when an investigator presents the monkey with a single object called the sample, for example, a colored plastic box or a piece of scrap metal. The animal can displace the sample object to uncover a raisin reward, which guarantees that the animal has for a moment attended to the object. After a delay of a few seconds, the animal is presented with the original object and a new object in a two-choice test. To obtain the raisin reward, the monkey must choose the new object. (In principle, one could reward the animal for choosing either the new object or the old object. In either case, a correct choice would indicate that the animal has recognized the sample object as familiar.) By varying the interval between the

sample and the choice from a few seconds to many minutes, one can track the ability of the animal to retain newly acquired material. The same procedure is repeated many times at each delay interval to obtain a robust measure of memory ability. New objects are used on every trial.

In this simple visual test of object discrimination, a monkey remembers which of two objects is the correct one.

The second task requires the monkey to learn and remember which of two simple objects is the correct one. In this task, the animal encounters on each trial two easily distinguishable objects, one placed to its left and the other to its right. One of the two objects is designated the correct object, and choosing that object results in a raisin reward. The position of the correct object (to the monkey's left or right) varies randomly, so that the animal learns to choose between the two objects themselves. The spatial location of the objects is irrelevant. A normal monkey requires 10 to 20 trials to learn which object is correct.

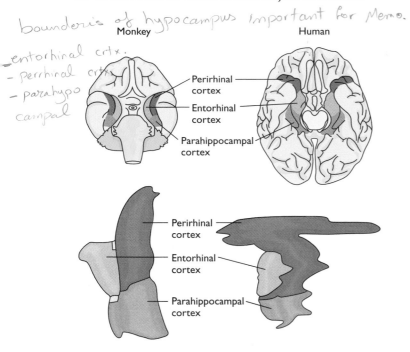

The medial temporal lobe. These views of the monkey and human brain from below, shown at the top of the figure, show the borders of the entorhinal, perirhinal, and parahippocampal cortices. In the human brain, the perirhinal cortex also extends alongside the entorhinal cortex, but this aspect of the perirhinal cortex is buried along the banks of a sulcus (i.e., a groove). Below them appear unfolded, two-dimensional maps of these cortical areas. These cortices, together with the hippocampus, dentate gyrus, and subiculum, make up the medial temporal lobe memory system on which declarative memory depends. The brains are not drawn to scale.

Work with monkeys by Zola and Squire at the University of California, San Diego, and by Mishkin and Elizabeth Murray at the National Institute of Mental Health, led to three important conclusions about the medial temporal lobe and memory. First, bilateral damage impairs memory, even when damage is limited to the hippocampal region. The impairment can be rather mild if the task is simply to recognize recently presented objects as familiar. Thus the monkey studies support the human studies in showing that the hippocampus is a component of the declarative memory system. Second, the amygdala is not a part of the declarative memory system. The amygdala is important for emotion and for aspects of emotional memory (as we shall see in Chapter 8), but it is not essential for declarative memory. Third, the cortex surrounding the hippocampus and amygdala is important for declarative memory.

The boundaries of this surrounding cortex and its connectivity to other areas became understood relatively recently. This cortex consists of three distinct areas: the entorhinal cortex, the perirhinal cortex, and the parahippocampal cortex. The major projections into the hippocampus itself originate in the entorhinal cortex. The entorhinal cortex in turn receives information from elsewhere in the cortex, approximately two-thirds of it from the adjacent perirhinal and parahippocampal cortex. All three cortical areas receive information from and send information to a broad extent of cortex. Thus, these areas have access to much of the processing that occurs in other cortical areas. But this cortex adjacent to the hippocampus is not simply a conduit for funneling information from other cortical areas into the hippocampus. Direct damage to the perirhinal and parahippocampal cortices impairs memory even more severely than damage to the hippocampal region itself. Thus, these cortical areas themselves contribute to declarative memory, and information need not reach the hippocampus

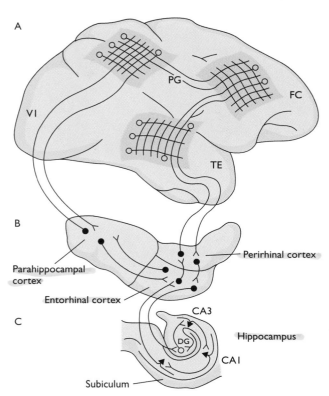

The pathways into and out of the medial temporal lobe memory system, shown here for the monkey brain, are believed to be important in the transition from perception to memory. For activity in areas TE and PG—which is influenced by activity in the frontal cortex (FC)—to develop into a stable long-term memory, neural activity must occur at the time of learning along projections from these regions to the medial temporal lobe—first to the parahippocampal cortex, perirhinal cortex, and entorhinal cortex and then in several stages through the hippocampus. The fully processed input eventually exits this circuit by way of the subiculum and entorhinal cortex and returns to areas TE and PG.

for some memory to be stored. In general, as more of the medial temporal lobe system is damaged, the memory impairment becomes more severe. However, this relationship does not mean that the medial temporal lobe structures all carry out a single function and that this function is gradually compromised as more and more of the region is damaged. The different structures of the medial temporal lobe may well carry out different subfunctions. According to this view, as the damage increases, fewer strategies remain by which memory can be stored.

PROPERTIES OF DECLARATIVE MEMORY

There is now good correspondence among the findings for all the well-studied mammalian species: rats, monkeys, and humans. In each species, an individual with damage to the hippocampus or to anatomically related structures will develop a circumscribed impairment in the ability to establish declarative memory. It also has become clearer how to characterize declarative memory in both humans and experimental animals. Declarative memory is well adapted for forming conjunctions (or associations) between two arbitrarily different stimuli. Some forms of declarative memory can be acquired rapidly, sometimes in a single trial. For example, a person can quickly learn to associate any two unrelated words (for example, "garden"–"jump" or "surprise"–"chime"). Other forms are acquired gradually, as when someone learns a long list of items or when a rat learns a spatial location. In either case, declarative memory is designed to represent objects and events in the external world and the relationships between them. A key feature of declarative memory is that the resulting representation is flexible. Animals can learn relations among stored items and then express this relational knowledge in novel situations.

The flexibility of declarative memory, and the relative inflexibility of nondeclarative memory, is nicely illustrated in a study of spatial learning and memory in the rat. Howard Eichenbaum and his colleagues at Boston University studied intact rats and rats with damage to the hippocampal

BEFORE LEARNING

Hidden platform

AFTER LEARNING

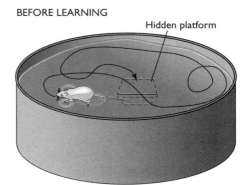

The Morris water maze. Left: The convoluted curved line shows the trajectory a rat or mouse might take to find a hidden platform the first time the animal is placed in the pool of murky water. Right: After training, the animal knows where the platform is and swims straight to it.

system. The animals learned to swim from the edge of a large circular pool of murky water to the location of a slightly submerged, hidden platform. Because the animals could climb onto the platform to escape from the water, finding the platform was in itself an effective reward. On each learning trial, rats were started from the same point along the circumference of the pool. Both groups learned the location of the hidden platform, as measured by sharp reductions in swimming time and reductions in swim length. Thus, as learning progressed, the rats came to swim straight for the platform location rather

than reaching the platform in a roundabout way. After learning was completed, the animals were made to perform additional test trials to determine what kind of information they had acquired about the location of the platform. In these trials, animals were started at novel locations around the circumference of the pool. Normal rats were able to find the platform quickly from any starting location, indicating that they had acquired a flexible (declarative) representation of space in memory. Specifically, they had learned about the spatial relationships between the platform location and various external cues that were available

SAMPLE

CHOICE

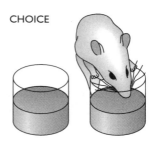

Odor-paired associates

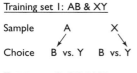

Training set 1: AB & XY

Sample	A	X
Choice	B vs. Y	B vs. Y

Training set 2: BC & YZ

Sample	B	Y
Choice	C vs. Z	C vs. Z

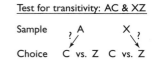

Test for transitivity: AC & XZ

Sample	A	X
Choice	C vs. Z	C vs. Z

Rats can learn a memory task that requires them to remember odors. Left: The rats first learn to associate a scent (the sample) with a buried reward (top), and then, when presented with two new odors, they learn which should be associated with the sample. Right: In learning this task, the rat first learns a series of associations (represented by the arrows) and then takes a test for transitivity that shows how flexibly it can use what it has learned.

around the walls outside the pool. By contrast, the rats with lesions were unable to find the platform from novel starting locations, and had to engage anew in a trial-and-error search around the pool.

The normal rats had acquired a declarative, relational kind of memory that could be accessed flexibly to guide behavior, even in new situations. In contrast, the animals with lesions had learned the same task, but they had learned a consistent relationship between specific cues and specific responses, a kind of nondeclarative, stimulus-response memory sometimes called habit learning that we shall encounter again in Chapter 9. An animal relying on habit memory can only retrace the same path on each succeeding trial.

The hippocampus of rats constructs a rich representation of space. As a result, there has been extended discussion concerning whether the hippocampus in rats is specialized for spatial memory (of the kind that would be needed to succeed in the swimming-pool task). To address this issue, Michael Bunsey and Eichenbaum tested rats with lesions limited to the hippocampus on a nonspatial task that involved learning associations between odors. After learning the associations, the rats were given additional tests to ask whether they could use the associations flexibly. Rats first learned to dig through a scented mixture of sand and rat chow to find a buried cereal reward. When the scent was cocoa, it signaled that on the very next trial the rats could find a reward buried in sand that was scented with coffee, but not in sand scented with salt. Thus they learned that following sample A, they should choose odor B (coffee), not odor Y (salt). They also learned three other problems of this same type. First they learned that following sample X, they should choose odor Y, not odor B. Then they learned that when odor B served as the sample, they should choose odor C, not odor Z. Finally, they learned that when odor Y served as the sample, they

should choose odor Z, not odor C. All the rats learned these associations.

Then, to test for flexibility, the investigators confronted the rats with a novel problem. The rats were now presented with odor A as a sample and given a choice between odor C and odor Z. Or they were presented with odor X as a sample and given a choice between odor C and odor Z. Intact rats chose odor C when presented A as a sample, and they chose odor Z when given X as a sample. In this way, the intact rats demonstrated their ability to infer relationships across stimulus pairs. Having learned, for exam-

Theodule Ribot (1839–1916), French psychologist who formulated the principle that injury to the brain impairs recently acquired memories more than remote memories. Archives of the History of American Psychology, University of Akron, Akron, Ohio.

ple, that A leads to B and that B leads to C, they "knew" that A also leads to C. However, rats with hippocampal lesions chose odors C and Z about equally often. Thus the intact rats used their memory flexibly, but rats with hippocampal lesions did not. In addition to confirming that animals can use hippocampus-dependent memory in new situations, this study provides definitive evidence that the hippocampus serves a general memory function in rats, just as it does in humans. Spatial memory tasks are best viewed as good examples of a larger category of declarative memory tasks that require the hippocampus.

THE TEMPORARY ROLE OF THE MEDIAL TEMPORAL LOBE SYSTEM

One of the remarkable features of declarative memory is that damage to the hippocampal system not only impairs new learning but also

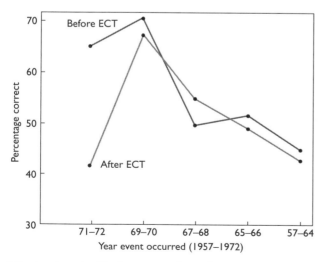

After treatment with electroconvulsive therapy, patients could not remember former television programs that they had seen one to two years previously. The test was given in 1974, and each program on the test had been broadcast for only one season from 1957 to 1972. Chance performance would equal 25 percent.

can disrupt some memories that were acquired before the damage occurred. This phenomenon, known as retrograde amnesia, was first studied carefully in the nineteenth century by the French psychologist and philosopher Theodule Ribot. Ribot observed that when brain injury or disease causes memory impairment, remote memory tends to be less affected than recent memory. His formulation of these observations came to be known as Ribot's law.

> This law, which I shall designate as the *law of regression or reversion* seems to me to be a natural conclusion from the observed facts. . . . This loss of memory is, as the mathematicians say, inversely as the time that has elapsed between any given incident and the fall [injury] . . . the new perishes before the old, the complex before the simple.

Memory is not fixed at the time of learning but takes considerable time to develop its permanent form. The fixation process requires several steps, one of which depends on the structures of the medial temporal lobe. Until the process is fully completed, memory remains vulnerable to disruption. Much of this process is completed during the first few hours after learning. But the process of stabilizing memory extends well beyond this point and involves continuous changes in the organization of long-term memory itself.

In the 1970s, a study of psychiatric patients by Squire established that it can take several years for a memory to become stabilized. These patients had all been prescribed electroconvulsive therapy for depression. Before and after treatment, the patients were tested for their memory of television programs that had broadcast for a single season during one of the preceding 16 years. Before treatment, the depressed patients remembered programs that had broadcast recently better than programs that had broadcast many years before. In other words, like most people, they remembered material learned recently better than material learned long ago. After treatment,

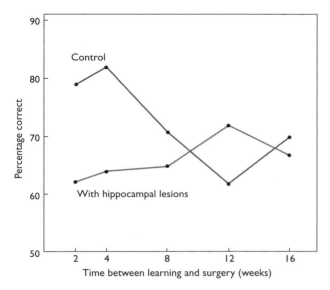

Two weeks after surgery to remove the hippocampal formation, monkeys had difficulty remembering recently learned objects, although their memory for objects learned many weeks ago was as accurate as that of unoperated monkeys. Chance performance would equal 50 percent correct.

the patients exhibited retrograde amnesia that followed a "temporal gradient": patients remembered the remote past normally (and better than the recent past), but their memory was poor for events from the previous one to three years. (These memories eventually returned.) The same point was subsequently established in studies of mice receiving electroconvulsive stimulation from one day to 10 weeks after a one-trial learning experience. The mice developed retrograde amnesia that covered about three weeks. Thus, the effects of electroconvulsive therapy show that memory grows gradually resistant to disruption over a relatively long time period. However, such studies could provide no information about what brain areas were important for the gradual stabilization of memory.

Since the time of Ribot, human amnesic patients have often been described as having relatively intact memory for events that occurred in their remote past. For example, patient H.M. was described as having a retrograde amnesia of several years together with intact memory for remote events. However, these observations, on their own, have been difficult to interpret because one is never sure how to compare recollections that come from different time periods. Questions about more remote time periods tend naturally to sample broader time intervals, to be more general, and even to ask about stronger memories than questions about the recent past. For example, you can ask someone what he or she ate yesterday, but when probing the more remote past your best option might be to ask about the highlights of a favorite holiday. Experimental animals provide a way to circumvent this problem. Animals can be given identical amounts of training at particular times before a treatment that affects memory. For this reason, studies with experimental animals have provided insights that even a hundred years of work with human patients could not provide.

Since 1990, several studies involving mice, rats, rabbits, and monkeys have found that retrograde amnesia occurs following damage to the hippocampus or anatomically related structures. In a study by Zola and Squire, monkeys learned 100 different pairs of objects prior to removal of the hippocampal formation bilaterally. One member of each pair was always designated the correct one, and monkeys learned for each pair to choose the correct object in order to obtain a raisin reward. Twenty pairs of objects were learned at each of five time periods before the operation (specifically, at 16, 12, 8, 4, and 2 weeks before surgery). After surgery, memory was tested by presenting all 100 object pairs once in a mixed order and allowing monkeys to choose the object that they remembered as the correct one. Only a single trial was given so that the measure of remote memory was not confounded with the difficulty animals would be expected to have in relearning the object pairs. The experiment showed that unoperated monkeys remembered recently learned objects better than objects learned many weeks earlier, as expected for normal memory.

hypocampal structures seem to be essential for learning and reorganizing the information.

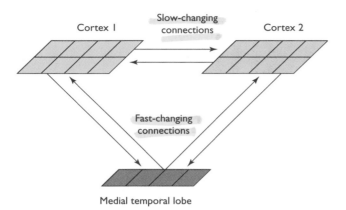

A model of how the storage of long-term memory might work. Each unit in each of the areas (four in the medial temporal lobe and eight in the two areas of cortex) is reciprocally connected to each unit in the other areas.

However, the findings were the opposite for monkeys with bilateral lesions of the hippocampal formation. Their memory of the remotely learned objects was normal, but they had trouble remembering the recently learned objects.

The hippocampal formation seems to be essential for only a limited period of time, a period that can range from days to years depending on what is being remembered. As time passes after learning, memory is reorganized and stabilized. During this period of reorganization, the role of the hippocampal formation gradually diminishes, and a more permanent memory is established, presumably in other cortical areas, that is independent of the hippocampal formation. James McClelland and Randy O'Reilly of Carnegie Mellon University and Bruce McNaughton of the University of Arizona have suggested that the fixation process allows these other cortical areas to change gradually, slowly incorporating into their representations facts about the world and other regularities of the environment. Thus cortical representations are not modified so rapidly that they become unstable and vulnerable to interference.

Do all declarative memories undergo this gradual fixation process? As described earlier, there had been a suggestion that spatial memories might have special status. Thus, the view that the hippocampus is important for spatial memory emphasized its role in learning and remembering places, including places learned about long ago. The idea was that spatial memories might always

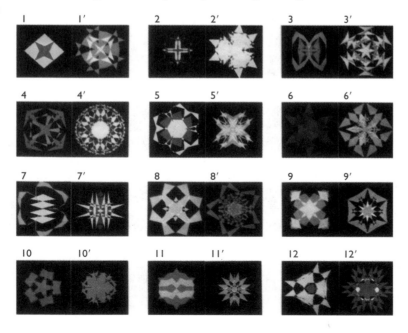

The 12 pairs of colored patterns used in a pair-association task developed by Sei-ichi Higuchi and Yasushi Miyashita to study declarative memory directly in the cortex of the monkey.

depend on the hippocampus, no matter how long ago the memories were acquired. However, it has turned out that spatial memories behave like other declarative memories. Edmond Teng and Squire found that even after complete damage to the hippocampus memory for places learned long ago is unaffected. They tested patient E.P., who was described in Chapter 1 as someone with large bilateral lesions of the medial temporal lobe and with no detectable ability to learn about new facts and events. Teng and Squire asked E.P. to recall the spatial layout of the region where he grew up, from which he moved away more than 50 years ago. On four different tests of spatial memory, E.P. performed as well as or better than five other individuals of the same age who had grown up in the same region and also moved away. In contrast, E.P. had no knowledge of his current neighborhood, to which he moved after he became amnesic. These observations show that the medial temporal lobe is not the permanent repository of spatial maps. The hippocampus and other structures in the medial temporal lobe are essential for the formation of long-term declarative memories, both spatial and nonspatial, but not for the retrieval of very remote memories, either spatial or nonspatial.

Information is not first stored in the hippocampal formation and then gradually transferred to areas of cortex outside the hippocampus. Memory is always in these cortical areas, but the memory system in the medial temporal lobe stores some aspect of the learned information for a significant period of time after learning. One possibility (illustrated on page 104) is that, after an event occurs, the medial temporal lobe rapidly stores links or pointers that connect it with the multiple cortical areas that together store a representation of the whole event. By this view, the medial temporal lobe is needed initially to support both storage and retrieval of the event, and it directs the gradual linking together in cortex of the neuronal ensembles that participate in the memory. Eventually, the network of

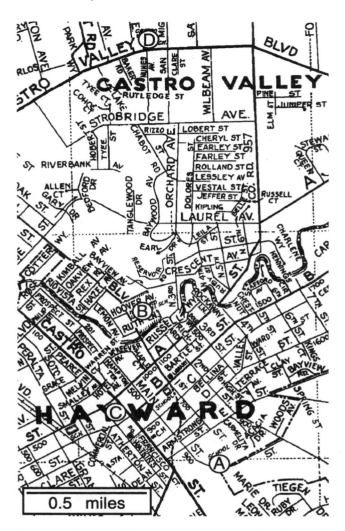

Street map from the 1940s shows a portion of the Hayward-Castro Valley region near San Francisco where amnesic patient E.P. grew up. Although he moved away more than 50 years ago, E.P. was able to recall the spatial layout of this region as well as others who grew up here. The letters indicate four of the landmarks used in the memory tests. A. Bret Harte School; B. Hayward Union High School; C. Hayward Theater; D. Castro Valley Grammar School.

interconnected cortical areas is able to support storage and retrieval without the help of medial temporal lobe structures.

The neural events underlying this gradual process of stabilization are not known. As we shall see in Chapter 6, however, an early step occurs within the hippocampus itself. Ultimately, long-term memory is thought to be stabilized by growth in the connections linking cortical areas. Depending on what is being learned, and the normal course of forgetting, the process of reorganization and stabilization can take days, months, or even years.

A study by Sei-ichi Higuchi and Miyashita in Tokyo raises the possibility of studying declarative memory in the cortex directly, as it develops and stabilizes in ensembles of neurons. Higuchi and Miyashita trained two monkeys to learn 12 different pairs of colored patterns, using the same procedure described on page 88. After training, they recorded from single neurons in the inferotemporal cortex while monkeys performed the task. As described earlier, with this procedure neurons can be found that "remember" the associations between the patterns. The scientists then recorded from these same neurons again, but after introducing lesions of the entorhinal and perirhinal cortices on one side of the brain, made shortly after training was completed. Neurons in inferotemporal cortex on the same side of the brain as the lesions no longer remembered the associations. The implication is that the sampled neurons were part of memory representations and that the adjacent medial temporal lobe needs to supply input to cortical areas in the inferior temporal cortex to maintain recently acquired representations in memory. This approach may provide a way to view directly the process by which the medial temporal lobe influences other areas of cortex. In addition, this work provides the first hint of an answer to a long-standing and fundamental question about memory. When memory is lost in amnesia after brain injury, are the memories truly lost from the brain, or can the loss potentially be reversed? Are the memories still present, but just inaccessible? If it proves to be the case that the neuronal responses observed by Higuchi and Miyashita do in fact reflect representations of declarative memory, then the loss of memory in amnesia reflects the actual loss of information from storage.

EPISODIC AND SEMANTIC MEMORY

So far we have been concerned with declarative memory for facts—for factual knowledge about objects, places, and odors. As early as 1972, the psychologist Endel Tulving used the term "semantic memory" to describe this kind of declarative memory for organized world knowledge. In recalling this type of information, an animal or human subject need not remember any particular past event. It need only know that, say, certain objects are familiar or that certain associations between odors are the correct ones. Semantic memory can be contrasted with *episodic memory*. As Tulving characterized it, episodic memory is autobiographical memory for the events of one's life. Episodic memory, unlike semantic memory, stores spatial and temporal landmarks that identify the particular time and place when an event occurred. For example, an episodic memory could involve the memory of going to dinner at a particular restaurant with a certain friend on a particular evening. Both episodic and semantic memory are declarative. Information is retrieved consciously and subjects are aware that they are accessing stored information.

The distinction between episodic memory (memory for particular times and places) and semantic memory (memory for facts) is useful. Semantic knowledge is thought to accumulate in cortical storage sites simply as a consequence of experience and support from the medial temporal lobe. In contrast, episodic memory is thought to require these cortical sites, in conjunction with the medial temporal lobes, to work together with the frontal lobes in order to store when and where a past experience occurred.

The role of the frontal lobes in episodic memory can be understood by considering the nature of episodic memory more closely. The essence of episodic memory is what is sometimes termed "source memory," that is, memory for where and when information was acquired. Impaired source memory is one consequence of impaired frontal lobe function, as shown by two kinds of evidence. First, source memory errors are rather common in young children and in the elderly, though not as common as after direct damage to the frontal lobes. This finding points to the frontal lobes as important for source memory, since it is known that they are slow to mature during development and are also compromised to some extent during normal aging. Second, patients with frontal lobe lesions tend to confuse where and when they learned what they know. A frontal lobe patient might remember from a recent learning session that "Cleo" was the name of the goldfish in the Pinocchio story, but then assert that he had learned this fact as a child or had heard it recently from a friend. Source remembering is at the heart of recollecting an individual episode from the past.

The frontal lobes are crucial for retaining source information and for maintaining the coherence of an episodic memory. When the content of some past event becomes disconnected from its original source, as in forgetting where one learned the name of the goldfish from the Pinocchio story, it may become connected to some other source or it may recombine with contents from other sources. The crucial role played by the frontal lobes in remembering source information provides one biological basis for some of the frailties and imperfections of declarative memory. Conceivably, even normal individual differences in the efficiency of declarative memory might reflect individual differences in the neuronal machinery of the frontal lobes.

The involvement of the frontal lobes in episodic memory has an interesting implication for the nature of learning and memory in nonhuman animals. Monkeys, rats, and other animals can clearly learn and remember a good deal. For example, they can remember "facts," such as the fact that choosing a red object will lead to a food reward. However, it is not at all clear to what extent nonhuman animals have the capacity for episodic memory, for example, the capacity to remember the earlier moment when choosing the red object resulted in a food reward. It has also been difficult to devise a satisfactory experiment to settle this question. In any case, it is worth considering the possibility that animals usually do not express memory for past events in the same way that people can—that is, as conscious autobiographical recollections of past happenings. Instead, animals (including rats and monkeys) may express memory mainly as currently available factual knowledge. Such a difference between humans and nonhuman animals would make sense in terms of brain organization. One of the striking differences in the brains of humans and nonhuman animals is the much greater size and complexity of the human association cortex, including the frontal lobes.

Consider the task of remembering a single event, such as the sight of a child playing with a toy airplane on the floor of a particular room. The frontal cortex exerts "top-down" control that biases neuronal activity in sensory cortex toward the relevant sensory information. This "top-down" influence, when directed across all the sensory cortical areas that are anatomically linked to the frontal cortex, would virtually define what is unique about the event. Antonio Damasio at the University of Iowa has proposed that much of recollection works in this way, with "top-down" activity from higher centers feeding back upon upstream cortical areas to re-evoke the specific features of an image or idea. The next chapter takes up a discussion of the cellular and synaptic mechanisms within the medial temporal lobe that are thought to store long-term declarative memory.

Pierre Bonnard, The Open Window *(1935). Bonnard (1867–1947), the Proust of French painters, was deeply influenced by the Impressionist and post-Impressionist artists who worked outdoors in order to capture the reflection of light onto objects. Years later, Bonnard was able to recreate these indoor and outdoor scenes and images from memory. In this scene he includes his wife as he remembers her during their early years together.*

6 | A Synaptic Storage Mechanism for Declarative Memory

Recall for a moment what you had for dinner the last time you ate at a restaurant and then try to recollect what wine you enjoyed with that meal. This effort at culinary memory requires your conscious recollection of declarative knowledge. By contrast, when you rushed to the net in last Sunday morning's tennis game and smashed the overhead lob hit by your opponent, you produced this sequence of motions unconsciously, and without any forethought, from your storehouse of nondeclarative knowledge.

We have already seen that declarative memory is concerned with conscious recollection of information about places, objects, and people, whereas nondeclarative memory is concerned with the unconscious performance of information about perceptual, motor, and cognitive skills and habits. Both types of memory are embedded in the particular sensory and motor systems that initially processed the

information. But whereas the learning of a nondeclarative memory tunes the efficiency of neurons
in these areas directly, the long-term storage of a
declarative memory requires an additional loop—
an extra system—namely, the hippocampus and
other structures in the medial temporal lobe.

Given the existence of these two very different memory systems, it becomes interesting to
ask: How different are the *basic* storage mechanisms? Does declarative memory with its requirement for conscious recall also require a specialized set of synaptic and molecular storage
mechanisms? We will address these questions in
this chapter and the next.

THE STORAGE OF DECLARATIVE MEMORY

As we saw in the last chapter, damage to the hippocampus in humans interferes with the storage
of new memories. To study these memory deficits
on a deeper, more mechanistic level, we need to
turn to experimental animals. Optimally, one
would like to study a comparatively simple experimental animal, but one that still exhibits the
capacity for declarative memory. Although experimental animals cannot declare anything about
what they remember, there are, as we saw in
Chapter 5, good reasons for believing that storage systems comparable to the declarative memory system of humans and monkeys also exist in
mice and other rodents. Mice and rats have many
of the characteristics of declarative memory that
are evident in humans. They can remember complex relationships among a number of different
cues in the environment, and they can remember
distinctions between objects. In particular, rodents form in the hippocampus a detailed internal
representation—a cognitive map—of space. As
we shall see, individual neurons of the hippocampus encode for space in their firing patterns. It is

thought that these characteristic firing patterns of
hippocampal neurons endow the animal with its
ability to remember a given space. Rodents require the hippocampus for spatial memory, for
memory of objects, and for other tasks characteristic of declarative memory but not for tasks that
depend on nondeclarative memory.

As we also saw in the last chapter, damage to
the hippocampus or other components of the medial temporal lobe system in humans does not interfere with memories that were stored long ago.
Patients like H.M. still have reasonably good
memory for the events that occurred earlier in
their lives. The situation is similar for experimental animals. Thus, the hippocampus is only a temporary storage site for long-term memory, for
periods ranging from days to months.

One way to view the role of the hippocampus
and the other components of the medial temporal
lobe is that they serve to modulate the initial representation established in cortical areas when the
information was first processed. According to this
view, the hippocampus serves a *binding* function.
It acts to bind together the storage sites that were
established independently in several cortical regions, so that ultimately these storage sites are
strongly connected with one another. We thus
need the medial temporal system for a lengthy,
but limited, period of time. The ultimate storage
site for long-term memory is thought to be in the
various areas of the cerebral cortex that initially
process information about people, places, and
objects. Most of what we so far know about the
long-term storage mechanisms for declarative
memory comes from studies of one region of the
medial temporal lobe system: the hippocampus.

TUNING SYNAPSES ARTIFICIALLY

In 1973 Tim Bliss and Terje Lømo working in Per
Andersen's laboratory in Oslo, Norway, made a
remarkable discovery. Aware of Brenda Milner's

insight about the role of the hippocampus and the medial temporal lobe in memory storage, they attempted to see whether the synapses between neurons in the hippocampus had the capability of storing information. To examine this possibility, they purposely carried out a quite artificial experiment. They stimulated a specific nerve pathway in the hippocampus of the rat and asked: Can neural activity affect synaptic strength in the hippocampus? They found that a brief high-frequency period of electrical activity (called a tetanus) ap-

plied artificially to a hippocampal pathway produced an increase in synaptic strength that lasted for hours in an anesthetized animal and would, if repeated, last for days and even weeks in an alert, freely moving animal. This type of facilitation is now called long-term facilitation, or more commonly, *long-term potentiation (LTP)*.

LTP has several features that make it suitable as a storage mechanism. First, it occurs within each of three principal pathways through which information flows in the hippocampus: the per-

The human hippocampus is a small structure, about the size of a child's thumb, located deep in the medial portion of the temporal lobe. Information flows into and through the hippocampus by means of the three principal pathways indicated in the cutout at the left: the perforant pathway, which runs from the entorhinal cortex to the granule cells of the dentate gyrus; the mossy fiber pathway, which runs from the granule cells of the dentate gyrus to the pyramidal cells of the CA3 region in the hippocampus; and the Schaffer collateral pathway, which runs from the CA3 region to the CA1 region.

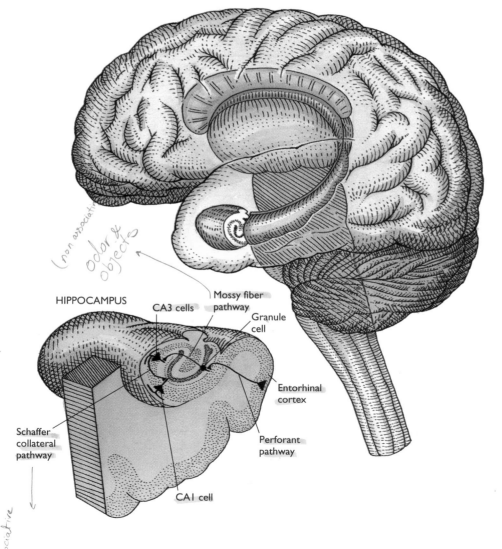

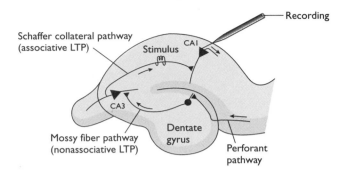

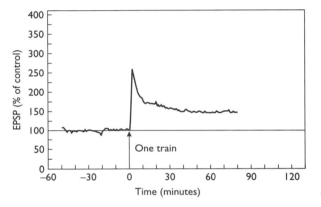

Long-term potentiation (LTP) as recorded in the Schaffer collateral pathway from the CA3 to the CA1 region of the hippocampus. Top: The experimental setup. A single train of electrical stimuli is applied to the Schaffer collateral pathway for one second at 100 Hz (100 impulses per second), and a microelectrode records the excitatory postsynaptic potentials (EPSPs) produced by a population of CA1 cells. Bottom: This single train of stimuli given at the arrow increases the strength of the synaptic connection between the CA3 and CA1 neurons, as measured by the EPSPs, for more than one hour.

forant pathway, the mossy fiber pathway, and the Schaffer collateral pathway. Second, it is rapidly induced: a single, high-frequency train of electrical stimuli can double the strength of a synaptic connection. Third, once induced, it is stable for one or more hours or even, as we shall see in the next chapter, for days depending upon the number of times the tetanus is repeated. Thus, as is the case for long-term facilitation in *Aplysia,* con-

sidered in Chapter 3, LTP has features of the memory process itself. It can be formed quickly at appropriate synapses and it lasts a long time.

Simply because LTP has features in common with an ideal memory process does not prove that it is the mechanism used for memory storage in life. If it could be shown to play a real, causal role in memory, however, it would offer scientists a great opportunity to study the storage mechanisms for declarative memory. Whereas the neural activity that goes into producing a declarative memory in ordinary circumstances is difficult to study, LTP is produced in the laboratory in controlled circumstances, under conditions that make it much easier to unravel the molecular mechanisms of memory.

Although LTP can be induced at several synapses in the hippocampus and in many regions of the cerebral cortex, the mechanisms for the induction of LTP are not everywhere the same. Detailed studies indicate that these mechanisms are of at least two major types: nonassociative and associative.

LTP in the Mossy Fiber Pathway

The dentate gyrus receives information from the entorhinal cortex and conveys that information to the hippocampus by means of the granule cells. These cells send out axons by means of a fiber bundle called the *mossy fiber pathway* that terminate on the pyramidal neurons of the CA3 region of the hippocampus. The mossy fibers release glutamate as their transmitter.

LTP in mossy fibers has features in common with the facilitation that occurs during sensitization in the *Aplysia* sensory neurons. Like the long-term facilitation which contributes to sensitization, mossy fiber LTP is nonassociative. It does not depend on postsynaptic activity or other signals arriving close in time. This form of LTP depends only on a burst of brief, high-frequency neural activity in the presynaptic neurons and the consequent influx of calcium. This calcium influx into the presynaptic neuron in turn initiates a familiar series of steps. Specifically, the calcium activates a calcium-calmodulin-dependent (Type I) adenyl cyclase; that enzyme acts to increase the level of cAMP; and the cAMP in turn activates the cAMP-dependent protein kinase (PKA). As we have seen, the cAMP-dependent protein kinase adds phosphate groups to proteins and thereby activates some proteins and inhibits others.

In *Aplysia,* the serotonin released by interneurons can modulate neuron activity and lead to long-term facilitation. Mossy fiber LTP is similarly influenced by a modulatory input, although in the case of LTP the transmitter is norepinephrine. This input engages receptors that bind to the transmitter, and these receptors activate adenylyl cyclase, just as serotonin activates adenylyl cyclase in *Aplysia.*

From its critical position in the hippocampus, one might expect that inactivating LTP in the mossy fiber pathway, which influences the firing of pyramidal cells of the CA3 region, might impair spatial memory formation. Yet surprisingly, recent attempts to explore its role in spatial memory suggest that it plays at most a minor role. Eugene Brandon and Stan McKnight at the University of Washington in Seattle, and Rusiko Bourtchouladze, Yan-You Huang, and Kandel, found that mutant mice with a selective defect in mossy fiber LTP learn spatial and contextual tasks normally. Thus, learning about space and context may be mediated by hippocampal circuits other than the mossy fiber-CA3 connection. It remains possible that LTP in the mossy fiber system is important for the storage of other kinds of declarative memory such as information about objects or odors. Alternatively, learning

(CA3 − mossfibers) does not have much with spatial role.

and memory may require ordinary synaptic transmission in the mossy fiber system, but may not require LTP.

Although the role in memory of mossy fiber LTP is still unclear, there is a better correlation between declarative memory and LTP in another hippocampal pathway: the Schaffer collateral pathway.

LTP IN THE SCHAFFER COLLATERAL PATHWAY

glutamate user but NMDA receptor activation are nessacary

The pyramidal cells in the CA3 region of the hippocampus send out axons to cells in the CA1 region, forming the Schaffer collateral pathway. The terminals of the Schaffer collaterals also release glutamate as transmitter, but, in contrast to the mossy fiber system, LTP is not induced in the Schaffer collateral pathway unless the NMDA-type of glutamate receptor is activated in the postsynaptic cell. Thus this form of LTP is associative; it requires concomitant activity both pre- and postsynaptically. To understand the mechanism of LTP in detail, it is first necessary to understand how LTP is initiated and, once initiated, how it is maintained.

Through the efforts of Jeff Watkins and Graham Collingridge in Bristol, England, it soon became clear that glutamate, the transmitter used in the Schaffer collateral pathway, acts here not just on one but on at least two major species of glutamate receptor in the receiving cell, an NMDA receptor and a non-NMDA receptor. The NMDA receptor channel does not function under routine circumstances, for reasons first determined by Phillipe Ascher at the Ecole Normale Supérieure in Paris and by Mark Mayer and Gary Westbrook at the National Institutes of Health. They found that normally the channel's mouth is plugged by magnesium (Mg^{2+}) ions that can be

2 major glutamate receptor ⟨ NMDA / others

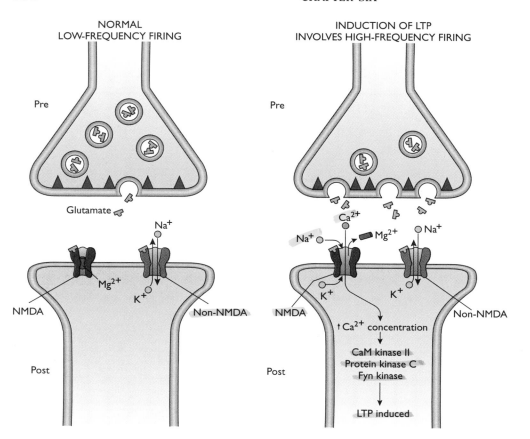

NORMAL
LOW-FREQUENCY FIRING

INDUCTION OF LTP
INVOLVES HIGH-FREQUENCY FIRING

The role of the NMDA receptor in the induction of LTP. Left: During normal synaptic transmission, when the presynaptic neuron fires at low frequency, the NMDA channels remain blocked by Mg^{2+} ions. Na^+ and K^+ ions can still enter through non-NMDA channels to mediate ordinary synaptic transmission. Right: LTP is induced when the presynaptic neuron fires at a high-frequency (a tetanus) and depolarizes the membrane of the postsynaptic cell sufficiently to unblock the NMDA receptor channel, allowing calcium to enter the cell.

displaced only when an especially strong signal is generated in the postsynaptic cell, a signal that reduces (or *de*polarizes) the resting potential of the post-synaptic membrane significantly. Such a strong depolarizing signal can be produced artificially by firing the presynaptic cells at a high frequency, and it is thought that a similar high-frequency burst of activity could also occur naturally during a learning experience. This strong firing reduces the membrane potential of the postsynaptic cell sufficiently to expel the Mg^{2+} plug from the mouth of the NMDA receptor, thereby allowing Ca^{2+} to flow into the postsynaptic cell through the NMDA receptor channel.

In a series of elegant experiments, Holger Wigström and Bengt Gustaffson in Göteborg, Sweden, placed these findings on the properties of the NMDA receptor into the context of LTP. They found that LTP requires not only that the presynaptic neuron fire but also that it fire repetitively so as to substantially depolarize the postsynaptic neurons and thereby remove the Mg^{2+} plug from the mouth of the NMDA receptor channel. Only then, they postulated, will sufficient Ca^{2+} enter through the NMDA receptor channels, thereby initiating the sequence of steps that leads to the persistent enhancement of synaptic transmission. This finding was interesting because it provided the first direct evidence

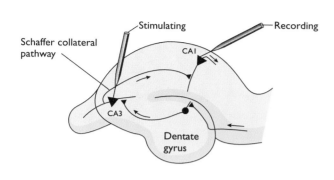

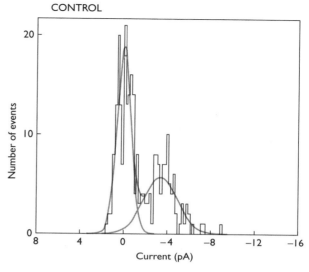

CONTROL

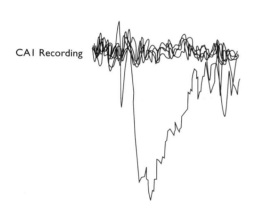

CA1 Recording

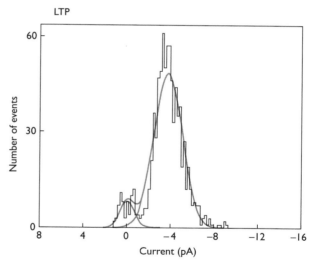

LTP

According to one view, LTP in the CA1 region of the hippocampus may depend not only on the insertion of new AMPA receptors postsynaptically, but also on an increase in presynaptic transmitter release. Using the experimental setup shown at the upper left, investigators recorded the current (measured in picoamperes, or pA) induced in a single CA1 cell in response to stimulation of a single presynaptic neuron in the CA3 region. In the control case, there were failures in transmission in the majority of cases, and relatively few successful responses, as indicated by the high curve centered around the 0 point and the low curve around 4 pA. Each inward current response of about −4 pA represents the release of a single quantum of transmitter. (Examples of failures and one unitary inward response are shown at the lower left). After LTP was induced, the percentage of failures decreased and the percentage of successes increased, an indication that the presynaptic cell was now more effective in releasing quanta of transmitter.

Influx of ca⁺ is crucial for induction of LTP

CHAPTER SIX

for Hebb's proposal, made in 1949, which stated that "when an axon of cell A . . . excites cell B and repeatedly or persistently takes part in firing it, some growth process or metabolic change takes place in one or both cells so that A's efficiency as one of the cells firing B is increased." Synapses that exhibit this property are now called Hebb synapses.

Soon after these steps were delineated, Gary Lynch at the University of California, Irvine, and Roger Nicoll at the University of California, San Francisco, provided direct evidence that the influx of Ca^{2+} through the NMDA receptor is the critical initiating signal for the induction of LTP.

The arriving calcium activates at least three protein kinases in the postsynaptic cell: the calcium-calmodulin-dependent protein kinase II, called CaM kinase II, protein kinase C, and a tyrosine kinase, fyn. Although these kinases are each different from the cAMP-dependent protein kinase we encountered in Chapter 3, they appear to serve an analogous function: they phosphorylate, adding phosphate groups to target proteins, and thereby activate some and turn others off. For example, Tom Soderling at the Vollum Institute in Oregon has found that the activated CaM kinase II phosphorylates the non-NMDA receptor in the postsynaptic cell, which then enhances the ability of these receptors to respond to glutamate released by the presynaptic neuron.

In addition, the work of Roberto Malinow at Cold Spring Harbor and that of Nicoll and Robert Malenka suggests that the action of CaM kinase II also influences the subsynaptic localization of AMPA receptors and results in new AMPA receptors being delivered and inserted into the synaptic membrane of the postsynaptic cell. In the extreme case some synapses may contain no AMPA receptors and have only NMDA receptors in their postsynaptic membrane. Because the NMDA receptor does not participate in routine synaptic transmission, these synapses are silent and ineffective prior to LTP. As new AMPA re-

ceptors are inserted into the postsynaptic membrane with LTP, these synapses become functional during routine synaptic transmission.

There is also reason to believe that LTP leads to an enhanced activity in the presynaptic neuron. This notion is based on the surprising discovery of Bliss, Charles Stevens at the Salk Institute, Richard Tsien at Stanford, and Steven Siegelbaum at Columbia, that following the postsynaptic induction of LTD there is an increase in transmitter release from the presynaptic terminals.

For example, in independent experiments Stevens and his colleagues, as well as Siegelbaum and Vadim Bolshakov, found that under normal circumstances a single CA3 neuron makes only a single synaptic contact on any one CA1 neuron. This single synaptic contact has only a single active zone from which it releases a single vesicle (containing the usual 5000 molecules of glutamate) in an all-or-none way. Before the induction of LTP each of these individual synaptic connections is quite ineffective: most action potentials in the presynaptic neuron do not succeed in releasing even this single vesicle. Therefore they produce no synaptic potential in the postsynaptic cell. By contrast, after the induction of LTP, most action potentials are successful in causing a single vesicle to be released and in producing a synaptic potential. Thus, LTP provides a vivid demonstration of how a synapse can be strengthened by activity. The most straightforward interpretation of this result is that at least in part, in its early phase, LTP results from an increase in the probability that a transmitter vesicle will be released.

Because the induction of LTP requires a postsynaptic event (the activation of NMDA receptors and a Ca^{2+} influx) and the maintenance of LTP appears to involve not only the insertion of new AMPA receptors into the postsynaptic cell but also a presynaptic event (an increase in the probability of transmitter release), it would appear that a *retrograde* message must be sent from the postsynaptic to the presynaptic neuron.

A New Type of Nerve Cell Communication

Because the induction of LTP requires a post-synaptic event (the activation of NMDA receptors and a Ca^{2+} influx) and the maintenance of LTP appears to involve, at least in part, a pre-synaptic event (an increase in the probability of transmitter release), it would appear that a message must be sent from the postsynaptic neuron to the presynaptic neuron. This is quite a radical idea. Since Ramón y Cajal first enunciated the principle of dynamic polarization at the turn of the century, every chemical synapse studied has proven to be unidirectional—information flows only from the presynaptic to the postsynaptic cell. However, LTP in the Schaeffer collateral pathway may require an additional mechanism, one that reflects a new principle of nerve cell communication. In response to second-messenger pathways activated by a calcium influx, the postsynaptic cell may release a signal that diffuses back to the presynaptic terminals, where it acts to enhance the probability that an action potential will trigger transmitter release.

What might be the nature of this retrograde signal? How does it work? The postsynaptic spines lack the release machinery of presynaptic terminals. There are no synaptic vesicles or active zones. It is therefore attractive to think that the retrograde messenger may be a substance that is synthesized as needed, rather than stored in vesicles, and that, once synthesized, readily diffuses out of the postsynaptic cell across the synaptic cleft and into the presynaptic terminal.

There is now evidence from the work of Dan Madison and Erin Schuman at the California Institute of Technology, and from Robert Hawkins, Ottavio Arancio, Kandel, and their colleagues, that there may be several retrograde messengers that work in this way. Of these, the one for which there appears to be the strongest evidence is nitric

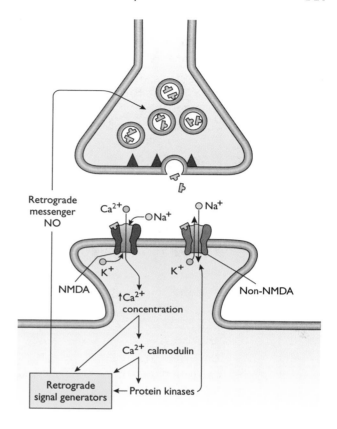

One view of LTP is that it works through a new type of nerve cell communication. According to this view, when NMDA channels are opened in the postsynaptic cell at the time LTP is induced, the resulting Ca^{2+} inflow into the postsynaptic cell and the resulting activation of CaM Kinase II and other protein kinases is thought not only to change the postsynaptic cell by acting on AMPA receptors, but also to send one or more retrograde messages back to the presynaptic cell telling it to release more transmitter. One of the retrograde messages is believed to be nitric oxide (NO).

oxide (NO). Nitric oxide is a fascinating example of a new class of messenger. It is a gas generated from the amino acid *l*-arginine by the action of the enzyme NO synthase. Nitric oxide can diffuse only a few cell diameters. Thus, although it moves freely, its distance of action is limited. Hawkins, Arancio, and Kandel have found that when nitric oxide is released from the postsynaptic cell, it enhances transmitter release if, and only if, it arrives in time to coincide with activity in the presynaptic neuron. In this respect it resembles activity-dependent presynaptic facilitation, which contributes to classical conditioning in *Aplysia*. Presynaptic activity appears to be critical for the ability of nitric oxide to maintain LTP.

What might be the advantage of combining two associative cellular mechanisms to produce LTP, the activation of the postsynaptic NMDA receptor and the recruitment of activity-dependent presynaptic facilitation? If presynaptic facilitation is produced by a diffusible substance, that substance could in theory find its way into neighboring pathways in addition to the one that was stimulated. In fact, studies by Tobias Bonhoeffer and his colleagues at the Max Planck Institute for Brain Research in Frankfurt indicate that LTP initiated in one postsynaptic cell spreads to neighboring postsynaptic cells. This enhancement in one or two adjacent cells could have the advantage of amplifying the effects of LTP. According

to this view, the unit for LTP would not be a single synapse but a small group of neighboring synapses. Because the retrograde messenger acts only on active presynaptic terminals, the action of the retrograde signal is not promiscuous. Not every synapse in the neighborhood would be potentiated, only those that are active at the moment.

LTP and Declarative Memory

LTP, as we have so far examined it, is a laboratory phenomenon, induced in a completely artificial manner. Thus, we cannot assume that it necessarily reflects what happens during the storage of a real memory. We therefore need to address two further questions: Does memory storage use LTP? If so, what precise role does LTP play? Here we shall address the first question, and in the next section we shall address the second.

If LTP is a mechanism for memory storage in the hippocampus, then defects in LTP should in-

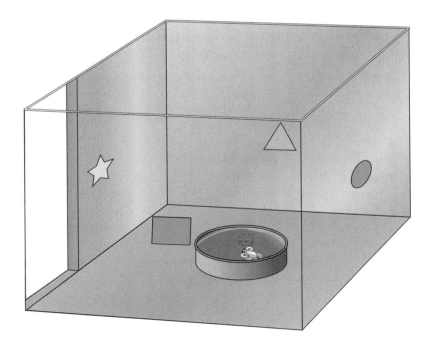

The Morris maze test requires a mouse to navigate its way to a hidden platform submerged in opaque fluid, as we first saw on page 100. A normal mouse can remember the location of the platform by using spatial cues on the walls of the room in which the chamber is located.

terfere with declarative memory. In the rat, as in humans, lesions of the hippocampus interfere with the formation of new spatial memories—memory for places, which is a form of declarative memory. Thus, one can ask: Is LTP necessary for the storage of spatial memories? Can one disable LTP and still have storage of spatial memories?

The first test of this question was undertaken by Richard Morris and his colleagues at the University of Edinburgh in Scotland. As we saw in Chapter 5, Morris developed a test for spatial memory that requires a rat (or mouse) to swim in a circular pool to find a platform submerged and hidden under an opaque fluid. The animal is released at a random location around the edge of the pool. On the first trial the rat eventually finds the platform by chance. To remember where the platform is on following trials, the animal must use the spatial cues provided by markings on the four walls of the room in which the pool is located. To use these cues, the animal needs declarative memory and the use of its hippocampus. By contrast, in a simple *nonspatial* (and nondeclarative) version of this test, the platform is marked with a flag so that it is visible directly. For this test, the mouse can navigate to the platform simply by heading toward the flag.

To test whether an NMDA-dependent form of LTP is necessary for the spatial task, Morris injected into the hippocampus an inhibitor that blocks NMDA receptors. With LTP blocked in this fashion, animals can navigate the *nonspatial* version of the task, in which the platform is marked, but they fail in the *spatial* version. These experiments suggest that, in the hippocampus, some mechanism of synaptic plasticity dependent on NMDA receptors, perhaps LTP, is involved in spatial learning and declarative memory.

There is a problem with relying on inhibitors to analyze a behavior or a biochemical pathway because inhibitors may not be completely specific. For example, they may also block other receptors or act on other molecules, and this other action may be the basis for their effect. Accordingly, research on memory took a major turn in 1990 with the development of gene knockout methodology. This technique allowed investigators to manipulate any gene in the genome of mice. In this way, it became possible to explore how manipulating a single gene affects both LTP and memory.

Genes, which are made of DNA, carry the blueprints for all the proteins that an organism can express, and they transmit this information from generation to generation through a process of replication. Thus, each gene provides succeeding generations (of mice and humans) with copies of itself. A particular gene directs the manufacture of a specific protein that helps determine the structure, function, or other biological characteristics of each cell in which it is expressed.

Genetically modified mice come in two major varieties, termed "knockouts" and "transgenics." In knockout mice, the gene of interest is deleted in all cells of the body and is absent for the entire life of the animal. As a result, conventional gene knockouts sometimes lack flexibility and precision: the experimenter does not have the option of eliminating the gene's activity in only certain cells or at only certain times.

In transgenic mice, an additional gene—the transgene—is added to the mouse genome by microinjecting DNA into the egg. The transgene may be the wild-type (natural) version of a gene—in which case the gene product is overexpressed—or it may be a mutant version of the gene, designed to enhance or suppress the gene's natural function. The transgene carries with it an appropriate promoter element, a sequence of DNA that directs when (in time) and where (in the body or brain) the gene is expressed. By including the appropriate promoters, the scientist can study the effects of a genetic modification primarily in the hippocampus, for example, and not in the rest of the brain.

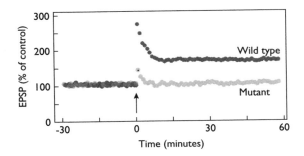

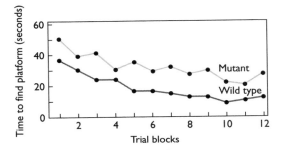

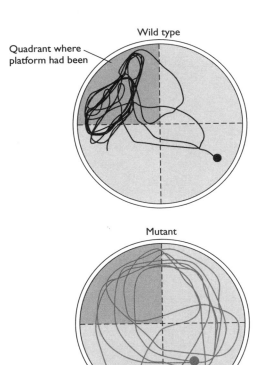

Mice generated by Joe Tsien and Susumu Tonegawa lack one subunit of the NMDA receptor only in the CA1 region of the hippocampus. These mice have a defect in LTP (top) and in spatial memory (bottom). Top: After a 30-minute period of baseline recording, a tetanus is applied for one second at 100 Hz. Activity in the CA1 region remains unchanged in the mutant mice, because LTP has been abolished. In contrast, LTP is induced in the wild-type mice. Bottom: Mutant mice are slower in learning to find a hidden platform in the Morris water maze by using spatial cues. The mutants show some improvement with practice, but never reach the optimal performance attained by the control mice.

After mice have been trained to perform the Morris maze, the platform is taken away. The lines in this idealized drawing indicate a typical swim path of a wild-type and a mutant mouse. Wild-type mice that have learned the task spend significantly more time than chance allows in the target quadrant, whereas mutant mice who have learned the task spend an equal amount of time in all quadrants. They do not show the normal ability to remember the location of the platform.

In 1992, Alcino Silva, Chuck Stevens, Susumu Tonegawa, and their colleagues at MIT and at the Salk Institute, and Seth Grant, Tom O'Dell, Paul Stein, Philip Soriano, and Kandel and their colleagues at Columbia University and at Baylor University, applied knockout techniques to study LTP and spatial learning in mice. They found that animals lacking one or the other of two second-messenger kinases—suggested by pharmacological experiments to be important in LTP—also exhibited reduced LTP. Knocking out these genes did not interfere with the normal behavior of these mice in any obvious way, so it was also possible to test the ability of these animals to learn and remember. In this way it was found that interfering with these two kinases also interfered with spatial memory. The mice lost their way in the spatial maze even after many training trials.

Even more direct evidence that LTP plays a role in spatial memory has come from studies of

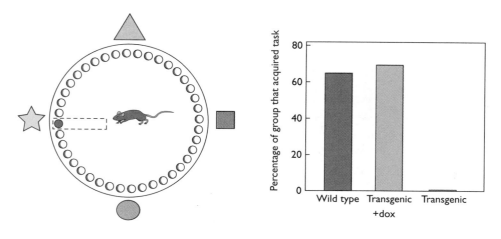

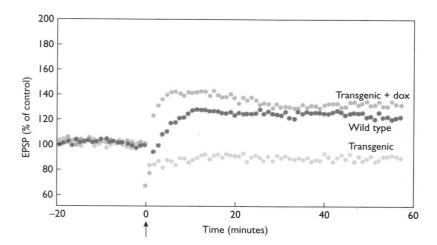

Mice that express the CaM kinase II transgene, which interferes with LTP and spatial memory, have normal LTP and normal spatial memory when the transgene is turned off by the drug doxycycline. Top left: A mouse is placed in the center of a Barnes maze consisting of a platform with 40 holes. One of these holes, shaded here for the reader, leads to an escape tunnel that is the mouse's only route of exit from this exposed and well-lit space. As with the water maze, the mouse's most efficient way of finding the hole is to use distinctive markings present on the four walls of the room in which the maze is located. The hole that leads to the escape tunnel is normally not marked. Top right: Transgenic mice that receive doxycycline perform the task as well as wild-type mice, whereas those without the doxycycline did not learn the task. Bottom: Recordings of excitatory postsynaptic potentials (EPSPs) show that stimulation at 10 Hz for 1.5 minutes induces a transient depression in response, followed by a modest amount of LTP, in wild-type mice, but only a slight depression in transgenic mice. Injecting doxycycline eliminates the defect in the transgenic mice.

RESTRICTING GENE KNOCKOUT

In their analysis of learning, biologists attempt to establish a causal relation that connects learning to the action of specific molecules. This relationship has been difficult in the past to demonstrate in mammals, but can now be studied more effectively in mice either by the selective knockout of genes or by the use of transgenes. In gene knockout, specific genes are deleted from embryonic stem cells through a process called homologous recombination, a genetic technique that was developed for use in the mouse by Mario Cappechi of the Howard Hughes Medical Institute at the University of Utah and by Oliver Smythies at the University of Toronto, Canada (Chapter 1).

As we have seen with conventional gene knockout techniques animals inherit the genetic deletions in all their cell types. These global gene knockouts make it difficult to attribute abnormalities to a particular type of cell within the brain.

To improve the utility of gene knockout technology, methods have been developed that restrict gene expression to specific regions. One method is to exploit the *Cre/loxP* system. Suppose you want to knock out the gene encoding one subunit of the NMDA receptor *(NMDA R1)*, but only in the CA1 region of the hippocampus. You need two sets of mice. One set will have been bred by conventional techniques to have two copies of the *NMDA R1* gene, each flanked by two *loxP* "recognition sequences." The *loxP* are short sequences of DNA that will be recognized by an enzyme called Cre recombinase. This enzyme recombines DNA strands in between the *loxP* sequences. The second set of mice carry as a transgene the *Cre* gene under the control of a promoter (in this case, *CaMKII*). For reasons that are still not understood, when the *Cre* gene is put under the control of *CaMKII* promoter, it sometimes effectively restricts recombination to the CA1 region of the hippocampus, perhaps because the transgene produces Cre recombinase at sufficiently high levels to initiate recombination only in the CA1 region.

Now the two sets of mice are mated. There will be among the offspring some mice that carry both the *Cre* transgene and the *NMDA R1* gene flanked by *loxP*. In these mice, the Cre recombinase when expressed at high level will recombine DNA between *loxP* sequences, thereby snipping out the *NMDA R1* gene.

The ability to turn a transgene off and on gives an investigator additional flexibility. Moreover, it allows the investigator to exclude the possibility that

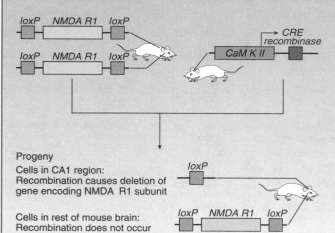

Transgenic mouse with two copies of NMDA receptor subunit gene flanked by *loxP* sites

Transgenic mouse in which *CRE recombinase* is expressed under the control of *CaM kinase* II promoter; CRE recombinase is effective only in CA1 cells

Progeny
Cells in CA1 region: Recombination causes deletion of gene encoding NMDA R1 subunit

Cells in rest of mouse brain: Recombination does not occur

The Cre/loxP *system for gene knockout. In this case, the gene targeted for knockout is the gene for the NMDA receptor. That receptor is composed of four subunits, so to incapacitate the receptor it is necessary to knock out the gene for only one of its subunits, called R1. The method shown here and in the figure on the facing page accomplishes gene knockout in a highly restricted manner, limited to the CA1 pyramidal cells of the hippocampus.*

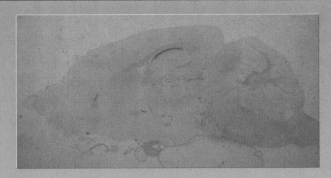

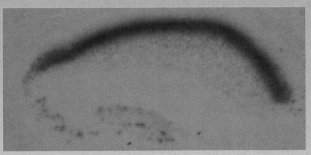

When the Cre transgene is driven by the promoter for the calcium-calmodulin-dependent kinase, the transgene is effective only in the CA1 region. This is evident in the two sections of a mouse brain shown here, one a low-power (top) and one a high-power (bottom) micrograph with a stain for ß galactosidase that reveals the action of the Cre recombinase. Thus the Cre recombinase has exerted its action only in the blue-colored layer of pyramidal cells that has absorbed the stain.

the abnormalities observed in mature animals are the result of a developmental defect. One strategy is to construct a gene that can be turned off by administering a drug. This was the strategy used to overexpress a form of calcium-calmodulin-dependent kinase that interferes with LTP and whose expression can be turned off and on. Again, one starts by creating two lines of mice. One line carries a transgene for calcium-calmodulin-dependent kinase (CaMKII), but instead of being attached to its normal promoter, it was attached to a promoter called *tet-o* that is ordinarily found only in bacteria. This promoter by itself cannot turn on the gene; it needs the help of a transcriptional regulator, and that help comes from a

transgene in the other set of mice. That transgene is the gene for a hybrid transcriptional regulator, called tetracycline transactivator (tTA), that recognizes and binds to the *tet-o* promoter. Expression of tTA is placed under the control of a promoter, the CaMKII promoter in this case. When the two sets of mice are mated, some of the progeny carry both transgenes. In these mice, the tTA binds to the *tet-o* promoter and activates the gene for the mutant calcium-calmodulin-dependent kinase, and this mutant kinase causes an abnormality in LTP. But when the animal is given the drug doxycycline, the drug binds to tTA and causes it to undergo a change in shape that makes it come off the promoter. The cell stops overexpressing calcium-calmodulin-dependent kinase, and LTP returns to normal.

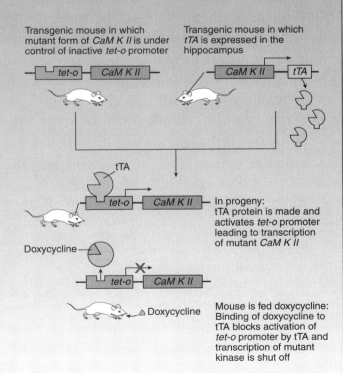

The strategy used to create a transgene that can be turned off by administering the drug doxycycline. In this case, the strategy is being used to control the expression of a constitutively active form of calcium-calmodulin-dependent kinase (CaMKII).

mice that have genetic lesions that interfere with LTP in a more restricted way. In one study Joe Tsien and Tonegawa at MIT selectively knocked out one subunit of the NMDA receptor in the pyramidal cells of the CA1 region. Although this disruption was restricted to the Schaffer collateral pathway, the mice nevertheless had a pronounced deficit in LTP in the CA1 region and a deficit in spatial memory. These findings provide compelling evidence that the NMDA receptor channels and LTP in the Schaffer collateral pathway are important for spatial memory.

However, gene knockouts, no matter how restricted, have a potential problem: because the defect is present from early in life, usually from birth onward, there could be developmental abnormalities in the mice. Thus, the defect in LTP and spatial memory could conceivably be the result of some developmental defect such as abnormal wiring of the Schaffer collateral pathway. This possibility was reduced through the use of a second type of mutant mouse generated by Mark Mayford and Kandel, in which a transgene could be turned on and off by giving a specific drug. In this second type of mouse, a mutated form of calcium-calmodulin-dependent kinase—one of the three kinases essential for LTP—was overexpressed throughout the hippocampus. Overexpression led to an interference with LTP and to a defect in spatial memory. However, when the transgene was turned off, LTP became normal and the animal's memory capability was restored. These experiments strengthen the correlation between some aspect of LTP within the Schaffer collateral pathway of the hippocampus and spatial memory.

These several findings raised the next set of questions: Why should interfering with LTP disturb the storage of spatial memories? How does LTP lead to the laying down of these memories? Only recently has it become clear that LTP seems to be required to form a stable internal representation of space in the hippocampus.

FORMING A STABLE MAP OF SPACE

In 1971 John O'Keefe and John Dostrovsky at University College, London, made the extraordinary discovery that the hippocampus can form an internal representation—a cognitive map—of its spatial environment. An animal's location in a particular space can be encoded in the firing pattern of its hippocampal pyramidal cells, the very cells that undergo LTP.

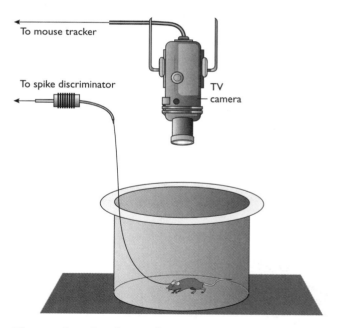

The recording chamber used to record the firing patterns of place cells. The head of a mouse inside the chamber is attached to a recording cable that is hooked up to a device able to resolve the action-potential firing ("spikes") from one or more CA1 pyramidal cells. As the mouse explores the chamber, the location of a light attached to its head is recorded by an overhead TV camera. Its output goes to a tracking device that detects the position of the mouse. The occurrence of spikes as a function of position is extracted from the basic data and used to form two-dimensional firing-rate patterns that can be either numerically analyzed or visualized as color-coded firing-rate maps.

WILD-TYPE MICE

TRANSGENIC MICE

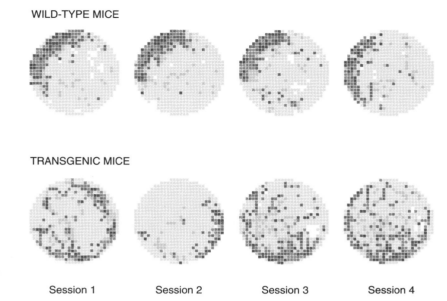

Session 1   Session 2   Session 3   Session 4

The firing-rate patterns from four successive recording sessions of a single place cell. Darker colors such as violet or red indicate high rates of firing and yellow indicates a zero firing rate. Before each recording session the animal was taken out of the circular enclosure and then reintroduced into it. During each session, the mouse explores equally all areas of the enclosure. However, each place cell fires only when the mouse is in a particular location. Every time the mouse is returned to the chamber, place cells fire when the animal walks through the same locations that fired those cells previously. The firing pattern for the cell from a wild-type mouse is stable. By contrast, the place field of a cell from a mouse carrying a gene for a persistently active calcium-calmodulin-dependent kinase is unstable.

A mouse's hippocampus has about a million pyramidal cells. Each of these cells can encode the features of the environment and the relationship among the features. One feature that they encode effectively is place. When the pyramidal cells encode information about place they are called, reasonably enough, *place cells*. When an animal moves around and enters different regions of a familiar environment, different place cells in the hippocampus fire. Some cells may fire only when the animal's head enters one position in a given space. Other cells will fire when the animal enters another position in the same space. Thus, a mouse's brain breaks up the space in which it walks into many small overlapping fields, and each field is assigned to a place in the hippocampus. By this means the animal is thought to form a spatial map of its surroundings. When the animal enters a new environment, new place fields are formed within minutes.

These observations have given rise to the idea that the hippocampus contains a maplike representation of the animal's current environment, and that the firing of place cells in the hippocampus signals the animal's moment-to-moment location within the environment. This spatial map is the best-understood example in the brain of a complex internal representation, a true cognitive map. It differs in several ways from the classical sensory maps found, for example, in the visual or somatosensory systems. Unlike sensory maps, the map of space is not topographic, that is, neighboring cells in the hippocampus do not represent neighboring regions in the environment. Moreover, the firing of place cells can persist after pertinent sensory cues are removed and even in the dark. Thus, although the activity of a place cell can be modulated by sensory input, activity is not dominated by sensory input as the activity of neurons in a sensory system is. It appears that the place cells do not map the current sensory input, but the location where the animal thinks it is in space.

Place fields are formed in minutes, and once formed, the map to which they contribute can remain stable for weeks. These facts raise several interesting questions: How do place cells contribute to memory for place? What are the mechanisms whereby place fields are formed and, once formed, retained? Is LTP important for the formation of place fields? For their maintenance? Do interventions that interfere with LTP and disrupt spatial memory do so by interfering with place cells and with the representation of space?

Studies of the two types of mutants we have just considered can address these questions, because each interferes with LTP in different ways. In one type, the NMDA receptors of the pyramidal cells of the CA1 region were selectively knocked out, causing a complete disruption of LTP in the Schaffer collateral pathway. In the other, a calcium-calmodulin-dependent kinase II was overexpressed throughout the hippocampus, interfering with LTP. In both types of mice, however, the place fields formed in a normal way when the mice were placed in a new environment.

Although LTP is not required for the formation of place cells, it is required for fine tuning their properties and specifically for the stability of the place field over time. In each mutant mouse,

the defects in LTP interfered with some specific properties of place cells. In normal mice, cells that have place fields near each other in space fire together in a relatively synchronous manner. When one cell fires, the other cell also tends to fire. T. J. McHugh, Matthew Wilson, and their colleagues at MIT found that this correlated firing is lost in mice lacking the NMDA receptor in the CA1 region. In mice that overexpress the calcium-calmodulin-dependent protein kinase, Alex Routenberg, Robert Muller, and their colleagues at the Downstate Medical Center and at Columbia found that the place fields are unstable over time. Whenever the animal is reintroduced into the same environment sometime later, the cells form different fields. This instability in place fields is thought to severely impair an animal's ability to learn and remember spatial tasks—information gained in a given training session is lost, and during a subsequent training session the animal behaves as though it is being presented with the task for the first time. If place cells are the building blocks of a cognitive map, the instability of place cells would make the map itself unstable, and therefore (after some time has passed) the map would no longer be suitable for making an efficient calculation of navigational paths. In fact, the consequences are very similar to the consequences of memory deficits seen in human patients with lesions of the medial temporal lobe. For patient H.M. each session of a multisession learning test is like the first: he does not remember that the experiment took place previously or even recognize the psychologists who administered the test.

Together these experiments provide an initial link in the chain of causation for creating a declarative memory that connects molecules to mind, by showing how genes affect the connections between cells and how these alterations affect an internal representation that guides a complex behavior of the animal. Specifically, interfering with LTP in the Schaffer collateral pathway of the hippocampus also interferes with the normal functioning of place fields—the internal representation of space. Defects in LTP interfere in particular with the stability of the spatial map over time. This unstable map in turn reveals itself in behavior as an unstable spatial memory.

ENHANCING LTP ENHANCES MEMORY STORAGE

As we have seen, the knockout of one of the subunits of the NMDA receptor by Joe Tsien and his colleagues interfered not only with LTP in the CA1 region, but also with spatial memory and with the hippocampal map for space, as evident in the place field map. Tsien, now at Princeton University, and his colleagues have also carried out the inverse experiment. They used the CaM kinase promoter to overexpress one of the subunits of the NMDA receptor, thereby allowing an extra amount of calcium to enter the CA1 pyramidal cells during the high frequency stimulation that induces LTP. They found that this increase in calcium influx into the postsynaptic cell led to enhanced LTP and to improved spatial memory and other forms of hippocampus-dependent memory, thereby producing a mouse wih a better memory for these tasks. We shall return to the issue of enhancing memory in Chapter 7.

SOME INTRIGUING SIMILARITIES BETWEEN DECLARATIVE AND NONDECLARATIVE MEMORY STORAGE

A striking finding is that the mechanisms for mossy fiber and Schaffer collateral LTP that we have encountered in the hippocampus are not unique to declarative memory but are used, with variations, time and time again. Thus the mechanisms for mossy fiber LTP are similar to those we encountered in Chapter 3 and which are used in *Aplysia* for the serotonin-mediated presynaptic facilitation that contributes to sensitization. Similarly, the NMDA receptor mechanisms for Schaffer collateral LTP resemble those of activity-dependent enhancement of presynaptic facilitation used in *Aplysia* for classical conditioning. These mechanisms may also come into play in the visual cortex, where they are thought to participate in the fine tuning of synaptic connections during the late state of development.

Thus, despite the logical and anatomical differences that distinguish declarative from nondeclarative memory, the basic short-term storage mechanisms used by these two memory systems share features in common. These similarities become even more pronounced as we consider, in the next chapter, the mechanisms used to convert short-term memory to long-term memory.

Roy Lichtenstein, The Melody Haunts My Reverie *(1965). Pop artist Lichtenstein (1923–1997) uses comic strip figures to parody contemporary culture while reminiscing about an earlier and simpler time reflected in the simplicity of a cartoon. Here he alludes to Hoagy Carmichael's beautiful song "Stardust."*

7 | From Short-Term Memory to Long-Term Memory

On the morning of August 31, 1997, Diana, Princess of Wales, lost her life in a senseless car accident in Paris while driving in the road tunnel under the Place de l'Alma. The driver, Henri Paul, and Diana's companion, Dodi al Fayed, died instantly. The Princess of Wales died several hours later in Pitié Salpétrière, a local Parisian hospital. The only person to survive was Trevor Rees-Jones, the princess's bodyguard. Badly hurt, his jaw shattered, he suffered a concussion that left him unconscious. Later examinations revealed that Henri Paul had been drinking and that his blood alcohol level was four times above the legal limit. But Paul's blood alcohol level may not have been the only cause of the accident. Francois La Vie, a witness close to the scene of the accident, told the French police that he saw the black Mercedes enter the tunnel surrounded by paparazzi, free-lance photographers on motorcycles, one of which happened to cut off the car before the crash. When the motorcycle cut in front of the car carrying Princess Diana, La Vie saw a large white flash from the camera of one of the photographers. According to La Vie, the flash might have blinded Henri Paul and contributed to the accident.

Rees-Jones was the only person traveling with Diana who could possibly testify to the events leading to the accident. But his concussion left him comatose for several days. When he finally regained consciousness, he recalled getting into the Mercedes and strapping on his seat belt, but he recalled nothing more of the events of the accident itself. We have every reason to believe that Rees-Jones had experienced and registered in his short-term memory all the events that occurred up to the moment of the crash. Yet after the crash, he remembered nothing about it. What caused this critical failure of memory, which prevented the only living witness from testifying to the facts surrounding this tragic accident? What was responsible for his amnesia?

Scientists have known for almost a century that the ability to convert short-term into long-term memory requires the throwing of a switch and that the processes that occur before the switch is thrown are readily disrupted, especially by brain injury. Thus, for Rees-Jones, the events that took place just before the accident never gained entry into a stable long-term memory because the switch did not function.

In a milder form, this sequence of events, including the resulting retrograde amnesia, is common and occurs frequently during the football season when a running back or a lineman is *dinged* after being hit particularly hard and, as a result, suffers a slight brain concussion. When a dinged player is asked about the play he had just participated in, *immediately* after being dinged, he may be quite dazed, but he nonetheless often remembers the name of the play and his part in it. But when asked *half an hour later*, the dinged player often no longer recalls the play or exactly what his part was, although now he is more fully recovered from the concussion and better oriented. Such instances show that the player clearly had the information in his short-term memory before the concussion, but that the blow to the head prevented the conversion of this short-term memory into a stable long-term form.

Consider a third, even more common experience—you try to recall the name of a person to whom you were recently introduced at a party. The ease with which you recall the person's name will depend on a number of factors: how interesting you found the person, how important the encounter was for you, how attentively you focused on the conversation, and what your general state of mind was that evening. As the experiences of everyday life illustrate, the ability to transfer information from short-term to long-term memory varies considerably from circumstance to circumstance. That is because the *switch* whereby short-term memory is converted to long-term memory is highly regulated so that the ease with which we convert short-term memory to long-term memory varies a great deal, certainly from day to day, but often even within a given day. In this chapter we shall first consider how long-term memory is stored and then return to examine the switch that initiates that long-term storage.

PRACTICE MAKES PERFECT

The study of long-term memory and its formation from short-term memory had its origins in the 1880s as part of the attempt by Hermann Ebbinghaus to create an experimental laboratory science for human memory. As was related in Chapter 1, Ebbinghaus devised simple techniques for measuring memory that are still in use today. He showed that what had appeared until then to be an intractable problem—the study of memory in human beings—could be attacked experimentally, provided one was willing to simplify the situation sufficiently.

Ebbinghaus's logic was simple. He wanted to study how new information is put into memory storage, but he had to be sure that the information learned by a subject was truly new. Thus he reasoned that he would need to ensure that a subject would have no, or at least minimal, old associations to material presented for learning. To force subjects to form only new associations, Ebbinghaus hit upon the idea of giving subjects lists of novel words—a new language so to speak. These words would be so totally unfamiliar that the subjects could not possibly have prior associations with them. Ebbinghaus invented the idea of using nonsense syllables as his novel words. These are syllables consisting of two consonants separated by a vowel (e.g., NEX, LAZ, JEK, ZUP, and RIF). Because a nonsense syllable is meaningless, it largely escapes the learner's preestablished network of associations. Ebbinghaus made up about 2300 such syllables and randomly used from 7 to 36 of them to create syllable lists varying in length. He then memorized each list of syllables by reading it aloud at the rate of 150 items per minute.

From these simple experiments, Ebbinghaus anticipated the distinction between short-term and long-term memory. He found that memory is graded in strength—that repetition is required to convert short-term memory to long-term memory. It is practice that makes perfect. By varying the number of repetitions during learning from 8 to 64, he found a nearly linear relationship between the number of repetitions used to learn the list and retention on the following day. Thus, long-term memory seemed to be a graded function of practice.

A few years later, Ebbinghaus's ideas were elaborated by William James, whose thinking about memory we also encountered in Chapter 1.

James concluded that there must be at least two different stages in memory storage. He proposed that there is a short-term process, which he called primary memory, and a long-term process, which he called secondary memory. Information that we have just acquired is consciously maintained in *primary memory* for a short period. By contrast, we summon *secondary memory* to actively call a memory *back* into conscious view some time after it has faded from consciousness following the original learning event.

As we saw in Chapter 5, James's primary memory is now called immediate memory, a term that refers to the information that occupies our current stream of thought. Immediate memory can be extended in time to last minutes or more by a rehearsal system called working memory. But even after rehearsal, the information remains in a transient form. This extended transient phase of memory, which can last as long as an hour or even more, constitutes what we call short-term memory. Short-term memory exhibits three features that shed light on its basic mechanisms of storage: (1) it is transient; (2) it does not require anatomical change to be retained; and (3) it does not require new protein synthesis. By contrast, James's secondary memory ultimately becomes what we now call long-term memory. Long-term memory can be stabilized by anatomical changes, as we will see, and these changes require new protein synthesis.

THE CONSOLIDATION SWITCH

James's suggestion that there are stages in memory storage was soon elaborated by Georg Müller and Alfons Pilzecker at the end of the nineteenth century. Using nonsense syllables similar to those used by Ebbinghaus, Müller and Pilzecker found that even after an event has been placed in memory, some time must pass for the memory trace to achieve a stable long-term form. During this period, which they called the *consolidation period,* memory is sensitive to disruption. Their key finding was that learning a second list of nonsense syllables immediately after learning a first list interferes with later recall of the first list. They called this effect *retroactive interference.* Subsequent studies with experimental animals and humans confirmed Müller and Pilzecker's observation that newly formed memories are susceptible to disruption, but without such disruption newly formed memories gradually become much more stable.

Müller and Pilzecker's finding immediately caught the attention of clinical neurologists. They had found that head injuries and brain concussions sustained in battle or in an accident can cause retrograde amnesia, a loss of memory for events that occurred just before the traumatic event. As we saw in Chapter 5, in some cases retrograde amnesia can extend backward for months or even years. However, as with Princess Diana's bodyguard, short-term memory for the events preceding an injury by minutes, or even hours, is especially vulnerable to retrograde amnesia. For example, a boxer who suffers a brain concussion may remember going to the sporting event itself and even climbing into the ring, but everything from then on will be blank. Undoubtedly, a number of events prior to the blow will have entered short-term memory—the opponent's movements during the early rounds, even the action leading up to the damaging punch itself and the attempt to avoid it—but the blow to the head will have occurred before any of these memory traces has had a chance to become consolidated. These memory traces fail to survive.

Beyond head injuries and concussions, a second clinical observation highlighted the differences between the early, labile phases and the later, more stable phases of memory: the appreciation that epileptic convulsions also give rise to retrograde amnesia. Epileptic patients are often amnesic for events immediately preceding a seizure, although the seizure has no effect on memory for earlier events. In 1901, the British psychologist William McDougall suggested that the retrograde amnesia of concussion and convulsion might be accounted for by a failure of consolidation, the concept introduced by Müller and Pilzecker. The first opportunity to explore rigorously the failure of the consolidation switch came in 1949 when C. P. Duncan produced electrically stimulated epileptic seizures in an experimental animal, the rat. Duncan found that with experimental animals, as with humans, the memory for a newly learned task is disrupted by a convulsive seizure most severely when that seizure occurs shortly after training, during the period of consolidation.

NEW PROTEIN SYNTHESIS FOR THE SWITCH TO LONG-TERM MEMORY

The first clue toward a biochemical understanding of the switch to long-term memory was encountered in 1963 when Louis Flexner, and subsequently Bernard Agranoff and his colleagues and Samuel Barondes and Larry Squire, observed that the formation of a long-term memory requires the making of new protein, whereas the formation of a short-term memory does not. In work by Barondes and Squire, a mouse was required to learn to turn left or right at the choice point of a T-shaped maze. Just before training, experimental animals received injections of a substance that inhibited protein synthesis (cycloheximide or anisomycin), while control animals received injections of saline. Both groups of animals learned the task perfectly well. Both groups also exhibited a perfectly good short-term memory when tested 15 minutes after training. However, animals that had received the inhibitor of protein synthesis had a profound loss of long-

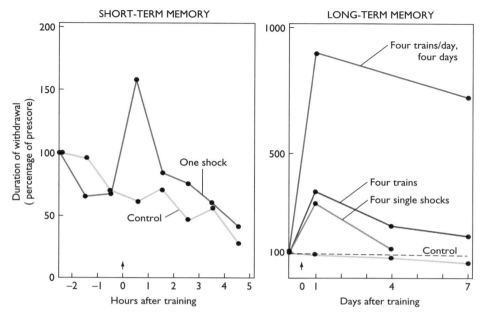

Before sensitization training, an Aplysia *will withdraw its siphon for about 10 seconds in response to mild touch. When an* Aplysia *receives a single mild stimulus to its siphon with a paintbrush every 30 minutes, it will habituate and withdraw its siphon for shorter periods of time. (Each point in the graph on the left represents the mean of two consecutive responses.) However, animals given a single tail shock (one shock after the third training trial to the siphon) withdraw their siphons for about twice as long in response to the paintbrush. The memory for short-term sensitization lasts about one hour (left panel). After receiving four tail shocks, the animal withdraws its siphon twice as long and now retains the memory for more than a day (right panel). If it receives four trains a day over the course of four days, the animal withdraws its siphon for almost eight times as long and retains the memory for more than several weeks.*

term memory when tested three hours or more after training. By contrast, control animals had perfectly good long-term memory. As with many other procedures that interfere with the formation of long-term memory, inhibitors of protein synthesis disrupt long-term memory only when given during a brief period usually of about one or two hours during and immediately after training. The inhibitors have no effect when given subsequently, several hours after training.

These experiments suggested that the consolidation switch to long-term memory requires new protein synthesis. What are these proteins? In the 1970s and 1980s the scientific methods were not available to delineate them in the mouse. The first

insights into these proteins came from studies of the marine snail *Aplysia*.

We have already encountered in Chapter 3 a form of long-term nondeclarative memory in *Aplysia*. A shock applied to the tail produces a sensitization of the gill withdrawal reflex: after a tail shock the animal responds more vigorously to a tactile stimulus to its siphon by withdrawing its gill farther. The duration of the memory of the tail shock depends on the number of repetitions of the tail shock. One shock to the tail produces a short-term memory lasting minutes. This short-term memory does not require the synthesis of new protein. By contrast, four to five shocks to the tail produce a long-term memory lasting one

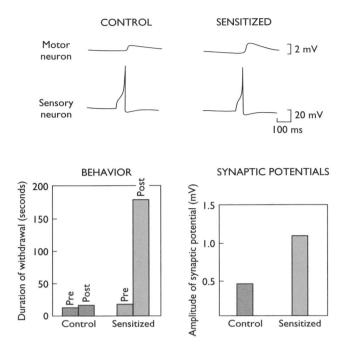

Evidence that synaptic connections between the sensory and motor neurons are enhanced with long-term sensitization. Top: One day after the end of training, the abdominal ganglion was exposed in both control and sensitized animals, and synaptic potentials were recorded in a siphon sensory neuron and a gill motor neuron. The synaptic connection from an animal that had received long-term sensitization training is stronger than that from a control animal in response to a presynaptic action potential of constant strength. Bottom: Quantification of the results shown above. The change in behavior (left) is paralleled by the change in synaptic potential (right). Tested after one day of training, the sensitized animals keep their siphons withdrawn for a considerably longer time than the control animals, and the connections from the sensory neurons to the motor neurons were twice as strong.

or more days, and further training produces an even more persistent memory lasting weeks. These longer-lasting changes do require new protein synthesis. It was soon revealed that this form of long-term memory displayed by *Aplysia* had as one of its components a strengthening of the connection between the sensory neurons and the motor neurons.

It is difficult to study the molecular mechanisms of the long-term change in an intact animal because even in a simple animal like *Aplysia* the investigator does not have good access, for long periods of time, to the cells involved in storing memory. But once scientists had identified these cells, it became possible to dissect them out of the animal's nervous system and place them in a culture dish. A major advance in the analysis of long-term memory was achieved in just this way, when Samuel Schacher and Steven Rayport at Columbia University succeeded in dissecting out from *Aplysia* individual siphon sensory neurons, gill motor neurons, and serotonin-releasing modulatory interneurons and explanting them into tissue culture. In effect, Schacher and Rayport reconstituted a component of the gill withdrawal reflex in a laboratory culture dish. In culture this highly reduced circuit exhibited many of the characteristic features that had been observed in the intact animal. In the intact animal, tail shocks excite the modulatory interneurons, causing them to release serotonin. In culture, as in the intact animal, these modulatory interneurons terminate on the sensory neuron, including on their presynaptic terminals, where they cause an increase in transmitter release from the sensory neuron. In fact, the serotonin-releasing interneurons could be dispensed with in tissue culture. One could simply apply serotonin directly to the sensory neuron by means of a puffer pipette that puffed the serotonin on the cell.

Investigators could now simulate the effects of repeated shocks to the tail by applying brief, repeated pulses of serotonin to the sensory neuron. In this way, Piergiorgio Montarolo, Castellucci, Schacher, Philip Goelet, and Eric Kandel found that, in tissue culture, the duration of synaptic facilitation goes up with the number of times serotonin is applied, much as is the case with the duration of sensitization in the intact animal. A single brief application of serotonin enhances the release of glutamate from the sensory neurons for several minutes. This short-term facilitation is not affected by inhibitors of protein synthesis and thus does not require the synthesis of new protein. By contrast, applying serotonin five times, at 20-minute intervals, enhances glutamate release for more than one day, and this long-term process does require new protein synthesis.

Earlier studies had shown that new protein synthesis was necessary to see a long-term change in an intact animal's behavior—in particular, to see the long-term sensitization of the gill withdrawal reflex. These cellular studies in culture showed that the same requirement existed at the synaptic level. In both simple and complex neuronal systems, protein synthesis is essential for long-term changes in synaptic connectivity. Just as long-term sensitization of the gill withdrawal reflex in the intact *Aplysia* requires the synthesis of new protein, so does long-term facilitation of the connection between the sensory and motor neuron of the gill withdrawal reflex.

Indeed, as with behavioral memory, long-term facilitation develops only if new protein is synthesized during a specific time window—during and just after the period of application of five pulses of serotonin. Inhibitors of protein synthesis block long-term facilitation only when applied during the application of serotonin or the subsequent one hour. When applied two to three hours after the first of the five applications of serotonin, inhibitors of protein synthesis no longer have any effect.

In the course of studying development, biologists had previously encountered cells that required new proteins rapidly and transiently during a discrete time period of two or three hours. In each of these contexts it turned out that genes had to be switched on. These findings therefore suggested that specific genes must be switched on for long-term memory to become established. The next logical questions were: What are the genes and proteins required to convert short-term memory to long-term memory and how are they activated?

SWITCHING GENES ON AND OFF

Nowhere has the ability of biology to delineate life's processes been evident more profoundly than in our understanding of what genes are and how they function. This understanding has given rise to the field of molecular genetics, which now forms the basis of much of biology. In fact, one of the reasons long-term memory is so interesting to the biologist is that the problem of long-term memory connects the cognitive psychology of a mental process—memory—to molecular genetics, the very core of modern biology.

Genes are stretches of DNA with two special properties that endow them with a unique double function. The first is that they can replicate. As a result, they can serve as templates—as repositories of genetic information that can be passed on from one generation to the next. The second property is that they encode the information needed to produce the proteins on which all aspects of life, including mental life, depends. This capability allows genes to serve as regulators of cellular function. Here, we shall be concerned primarily with the second function of genes.

With rare exceptions, every cell in the human body contains precisely the same complement of genes as every other cell (about 80,000 genes in humans). The reason that cells differ from one another—that a kidney cell is a kidney cell and a neuron a neuron—is that each type of cell expresses a different combination of the genes in its nucleus. This *differential* or *selective expression* of the genes underlies all cellular specialization. Thus there must be mechanisms for activating (turning on) some genes while repressing (turning off) others. Differential gene expression acts to create the defining features of nerve cells. The approximately 10^{11} neurons that make up the brain are divided into about 100 cell types that can be distinguished by their shape and their connections, features determined both by the distinctive combination of genes expressed within each cell and by the combination of genes expressed in the target cells with which each cell type interacts.

Which genes a given nerve cell will express determines its fate, the type of cell it will become as a functioning part of the brain. This decision is made during an animal's early development. In a given cell, a particular pattern of gene repression and activation occurs during development and then is typically maintained throughout the mature life of the cell.

Most of the genes that are expressed in a cell are transcribed in the nucleus into a molecule called *messenger RNA*. The messenger RNA molecule is transported out of the nucleus into the cytoplasm of the cell, where it is translated into a protein by the ribosomes, the protein-synthesizing machinery of the cytoplasm. The function of each messenger RNA is to carry the information about making a specific protein, which is encoded in the DNA sequence of the gene, from the gene in the nucleus to the ribosomes in the cytoplasm.

In a typical cell, 80 percent of the genes are repressed and only 20 percent are expressed (so that in a given cell a total of about 16,000 genes is expressed). In addition, all cells contain mechanisms that control the amount of protein produced from the 20 percent of active genes. Some of the transcribed genes are relatively inactive and express at low levels. These may produce less than 0.01 percent of all protein produced by the cell. Other genes are extremely active and may produce fully 10 percent of the total protein in the cell. Moreover, a gene's level of expression often is not fixed within a cell, but may vary over time. The most common way to regulate these rates is called *transcriptional control*. Transcriptional control mechanisms regulate the rate at

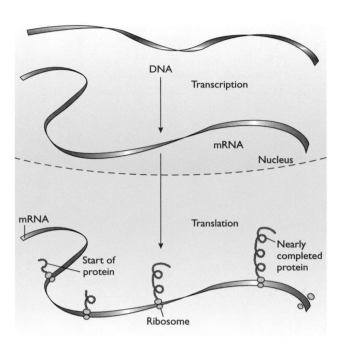

Transcription and translation are the two main steps in the creation of a protein from its DNA blueprint. The DNA is transcribed in the nucleus into a messenger RNA, which leaves the nucleus and is then translated on ribosomes into protein.

which a given gene is transcribed, when and to what degree it is turned on, and when and to what degree it is shut off.

How does all this relate to memory? The mechanisms whereby genes are turned on and off are critical to understanding how long-term memory is switched on and off. For example, serotonin must somehow regulate gene expression in the sensory neurons of the gill withdrawal circuit of *Aplysia* so as to establish new synaptic connections. Serotonin does so, as we shall see, by activating special regulatory protein molecules that can turn genes on and off.

How genes are switched on and off is a fascinating and important topic. Fully one-fifth of *all* genes in the human genome—about 16,000 genes—encode proteins, both activators and repressors, that regulate the expression of other genes! The discovery that genes can be regulated and the unraveling of the mechanism whereby the regulation is achieved is one of the most beautiful chapters of modern biology. The initial sentences of this chapter in modern biology were indelibly written by two great French biologists, Jacques Monod and Francois Jacob. Many years later, Jacob described their joint work in the following terms:

> For several years Jacques Monod and I spent hours everyday in his office in front of the blackboard, half talking, half sketching models. Progressively, . . . there emerged a representation . . . of the control unit of gene expression: a regulatory protein and its DNA target.

Let us first consider Monod and Jacob's target, which is a site on the DNA. Any given gene can be divided into two major regions: a coding region and a control or target region. The DNA in the *coding* region can be transcribed into a messenger RNA and then translated into a protein. Whether a given coding region is read out (transcribed) is determined by the pattern of regulatory proteins that binds to the control region. The *control* region usually lies *upstream,* just before the beginning of the coding region, and it is further divided into two subregions: the *promoter region* and the *regulatory region.* The regulatory region is further subdivided into 6 to 10 smaller regions called DNA *response elements.* Each of these response elements in turn recognizes and will bind specific *regulatory proteins.* These regulatory proteins are called *transcription regulators,* and they are of two sorts: activators that enhance transcription and repressors that shut it off. It is these transcription regulators that are essential for long-term memory.

Next door to the regulatory region, the promoter region binds proteins continuously, and this binding determines the steady-state level of transcription in the absence of stimulation. By contrast, the various response elements typically bind regulatory proteins only intermittently, as activators and repressors arrive to activate

Upstream from a gene's coding region are the regulatory and promoter regions that control the initiation of gene transcription. The transcription of DNA into mRNA is handled by a polymerase enzyme that initially binds to regulatory proteins attached to both the promoter and regulatory regions. Proteins that bind to the regulatory region can cause the DNA to loop so that these proteins can contact the polymerase enzyme. If no regulatory proteins are bound to the control regions, or if repressors have bound in their place, the gene cannot be transcribed.

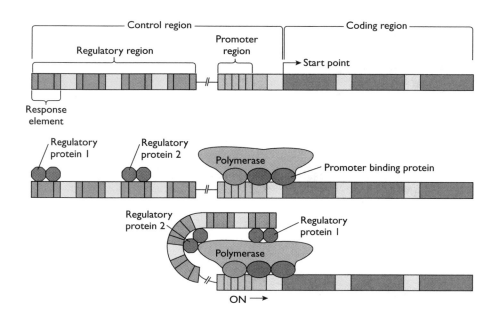

(induce) or repress (shut off) the gene and then depart once they are no longer needed. Here we meet up again with second-messenger systems, for it is the function of the appropriate second-messenger systems to signal the appropriate transcriptional regulators to bind to their different response elements, the DNA targets of Monod and Jacob. Thus, whether a gene is transcribed and how often it is transcribed in any given period of time is determined by transcriptional regulators that bind to the different response elements of the regulatory region. Thus, for serotonin to induce gene expression, it needs first to activate specific transcription factors that can then bind to specific response elements in the control region of genes important for long-term memory.

The idea that the transition from short-term to long-term memory storage requires genes to be turned on has connected the emerging molecular biology of memory storage to the well-established biology of gene regulation. In this way, it became possible to approach experimentally the question posed earlier in this chapter: What genes and proteins are required for long-term memory?

A NEW CLASS OF SYNAPTIC ACTIONS

To discover the genes and proteins required for long-term memory, it is necessary to return again to examine the properties of the cAMP-dependent protein kinase (PKA). As we saw in Chapter 3, cAMP-dependent kinase protein is a tetramer made up of four subunits. There are two catalytic subunits, each of which serves as the catalytically active enzyme able to phosphorylate proteins, and two regulatory subunits that inhibit the catalytic subunits. Only the regulatory subunits have binding sites for cAMP. When the level of cAMP rises, the regulatory subunits bind cAMP. As a re-

sult of this binding, the regulatory subunits undergo a shape change that frees the catalytic subunits. The freed catalytic subunits are then able to phosphorylate the target proteins and thereby alter their function.

Roger Tsien at the University of California, San Diego, developed an ingenious fluorescent labeling technique that allowed him to track the location of both the regulatory and the catalytic subunits of PKA within a single sensory neuron of *Aplysia*. Using this technique, Brian Bacskai and Tsien found that a single pulse of serotonin, which produces short-term facilitation, produces only a transient rise in cAMP. The catalytic subunits are freed up only for a period of minutes and primarily in the presynaptic terminal. Their action is responsible for the enhanced transmitter release that strengthens the connection of the sensory neuron to the motor neuron for a period of minutes. Following a single pulse of 5 HT, there is time for only a small number of molecules of the catalytic subunit to diffuse to the nucleus. With repeated pulses of serotonin, however, the level of cAMP rises to the point where the catalytic subunits are freed long enough so that a significant number will translocate to the nucleus. In the nucleus, the catalytic subunits activate genes that are critical for the growth of the new synaptic connections essential to long-term memory.

These findings show at the cellular level why repeated training trials or repeated pulses of serotonin are necessary for long-term memory. The repeated pulses allow the active component (the catalytic subunits) of protein kinases to travel to the nucleus, where they can activate genes necessary for the long-term process.

The discovery that PKA translocates to the nucleus provided a new perspective on synaptic transmission by revealing a new, third class of synaptic actions. As we saw in Chapter 3, Bernard Katz and Paul Fatt first described in 1951 the ionotropic receptors, receptors that control ion channels directly. They found that these

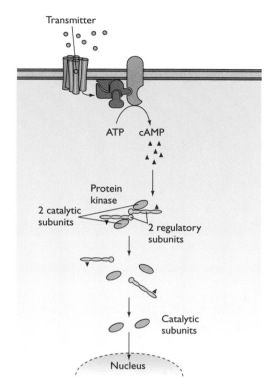

A chain of steps leading to the translocation of the catalytic (enzymatic) subunit of PKA from the cytoplasm to the nucleus of a nerve cell. The binding of transmitter to a metabotropic receptor engages the adenylyl cyclase enzyme. That enzyme transforms ATP to cAMP, and the cAMP removes the two (inhibitory) regulatory subunits from the catalytic subunit so that it may diffuse into the nucleus.

receptors mediate the common, fast synaptic actions, lasting only a few milliseconds, that are critical for mediating behavior. In the 1960s, Earl Sutherland, Paul Greengard, and their colleagues described a second class of receptors—the metabotropic receptors—that activate second-messenger pathways within the cell. This second class of receptors can initiate modulatory actions that last for minutes. This is the type of action that can, for example, modulate the strength of pre-existing synapses.

We now see that repeated sensitizing stimuli (such as tail shock) or repeated applications of

serotonin initiate a third type of synaptic action, an action that lasts not for seconds or minutes, but for days and weeks. This persistent action results from the fact that the repetition of the stimulus gives PKA opportunity to translocate to the nucleus. There it activates a genetic cascade, which then initiates, as we shall see below, a stable, self-maintained growth process within the neuron. Thus, modulatory transmitters such as serotonin that are important for learning may serve a double function. On the one hand, a single exposure produces a transient change in synaptic strength lasting minutes; on the other hand, a repeated or prolonged exposure produces a persistent, stable change in the architecture of the neuron.

FIRST STEPS IN THROWING THE CONSOLIDATION SWITCH

We have seen how the application of repeated pulses of serotonin sets off a chain of responses in the postsynaptic cell, all for the purpose of sending a signal to the nucleus. Now at last we can begin to see what happens inside the nucleus to create a long-term memory. After the PKA translocates to the nucleus, it phosphorylates a number of transcription factors, but perhaps the most important one is a protein called CREB-1 (the *cAMP-response element binding protein-1*). It is the action of the phosphorylated form of CREB-1 (bound to its *cAMP-response element* or *CRE* on the DNA) that switches on the genes needed to form long-term memory.

The first evidence that CREB-1 has a role in memory storage appeared in 1990 when Pramod Dash, Binyamin Hochner, and Kandel explored the question: What happens to long-term facilitation of the gill withdrawal response in *Aplysia* if CREB-1 is prevented from activating genes? To address this question, they synthesized a piece of DNA that encodes the sequence of *CRE*, the response element on the DNA to which CREB normally binds. They then injected this sequence in large amounts into the nucleus of an *Aplysia* sensory neuron. In so doing, they gave the CREB-1 protein a choice, as it were, of binding to the normal target *CRE* on genes or of binding to the large excess of *CRE* pieces of DNA that had been injected. Because there was much more injected *CRE* present than naturally occurring *CRE,* most of the CREB-1 protein become bound to the injected *CRE.* Dash and his colleagues thus succeeded in preventing the native CREB-1 protein from binding to the native *CRE.* In this way, the CREB-1 was rendered ineffective. The result was that the long-term process was blocked without interfering with short-term enhancement of synaptic strength.

Subsequently, Dusan Bartsch, Andrea Casadio, and Kandel carried out the reverse experiment. They injected the phosphorylated form of the CREB-1 protein into the sensory neurons of *Aplysia* and found that it produced long-term facilitation. It did so without producing the brief enhancement of transmitter release characteristic of short-term facilitation. Thus, in its phosphorylated form, CREB-1 is not only *necessary* for long-term facilitation, it is also *sufficient* to achieve it all on its own. These experiments provided direct evidence that CREB-1 is a critical component in the very first steps of the switch from short-term to long-term memory.

CONSTRAINTS ON LONG-TERM MEMORY

So far the molecular process of long-term memory storage can be characterized as follows. The activated catalytic subunit of PKA translocates to the nucleus and phosphorylates CREB-1 proteins. These proteins then bind to DNA and turn on genes needed to form long-term memory. However, the situation is not quite this simple. Per-

haps the most surprising complexity to emerge from studies of the genetic switch is that long-term memory can be switched off as well as on. Normally, the ability to establish long-term memory is actually constrained by inhibitory processes. These processes determine the ease or difficulty with which short-term memory will be converted to long-term memory.

The most dramatic constraint is an inhibitory transcription regulator—a repressor—discovered by Bartsch and Kandel called CREB-2. CREB-2 is thought to inhibit the actions of CREB-1, and thus block long-term facilitation, by binding to both the *CRE* target on the DNA and to CREB-1. Thus, to activate long-term facilitation in *Aplysia,* one has to not only activate CREB-1 but also remove the repressive influence of CREB-2.

CREB-2 is regulated in a different way from CREB-1. Unlike CREB-1, CREB-2 is not directly turned on by PKA. Rather, CREB-2 seems to be regulated by a different protein kinase—one called mitogen-activated protein kinase, or MAP kinase. This ability to regulate CREB-2 independently, and thereby to shut it off independently and to varying degrees, is interesting. It might account for some of the everyday variability that we experience in the ease with which we can convert short-term memory into long-term memory. If this argument is correct, then removing the repressive influence of CREB-2 should dramatically reduce the threshold for converting short-term to long-term memory. Mirella Ghirardi, Bartsch, and Kandel tested this idea and found that blocking the repressor gives rise to long-term synaptic facilitation important for long-term memory. Now a single exposure of serotonin, which normally produces short-term effects lasting only minutes, produced a growth of new synaptic connections lasting more than one day.

A similar form of coordinated regulation also controls the switch to long-term memory in the fruit fly *Drosophila.* As we saw in Chapters 1 and

3, Seymour Benzer and his students were the first to show that *Drosophila* can learn and that mutations in single genes interfere with the short-term memory. More recently, Tim Tully at Cold Spring Harbor Laboratories found that *Drosophila* also has a long-term memory, that it was activated by repeated training at spaced intervals, and that it was dependent upon new protein synthesis. Jerry Yin, Tully, and Chip Quinn went on to clone the two forms of CREB, one an activator and the other a repressor, in transgenic flies. They then overexpressed the CREB repressor and found that an abundance of the repressor blocked the formation of long-term memory. Perhaps even more striking, Yin and Tully found that overexpressing the activator form of *Drosophila* CREB greatly reduced the number of training trials needed to establish long-term memory. A single training trial that normally produced short-term memory lasting minutes now produced long-term memory lasting days. Together the data in *Drosophila* and *Aplysia* are complementary: they provide the molecular evidence that in addition to an activator important for switching on long-term memory, there is a special repressor that functions to prevent information from being put into long-term memory storage.

The fact that both activators and repressors have control of the consolidation switch may provide some novel insights into several features of memory, ranging from forgetfulness to feats of exceptional memory. For example, as we shall learn in Chapter 10, when one grows older one has difficulty in establishing long-term memories (age-related memory impairment), so one may not remember a recent conversation with a new acquaintance. This difficulty is thought to be due in part to a loss of synapses in the medial temporal lobe and in part to physiological changes that occur in that lobe with aging. However, the forgetfulness of normal aging could additionally be caused by a weakening in the stimulation from activators, and perhaps also to an inability to relieve repression. Moreover, people may show inborn differences in the activity of the repressor in relation to the activator. Thus our genetic endowment could contribute to individual differences in storage processes and account in part for people with exceptional memory.

EXCEPTIONAL MEMORY

Although we typically form a long-term memory only after repeated training at spaced intervals, occasionally new information becomes firmly fixed in mind following a single exposure. One-trial learning is particularly well developed in certain rare individuals with exceptional memory. For example, the famous memorist D. C. Shereshevskii (whom we learned about in Chapter 4), seemed never to forget anything he had learned, even following a single exposure, and even after more than a decade. More commonly, memorists are more restricted in their capabilities. For example, the Shass Pollaks, the Talmudic memorists from Poland, have a remarkable memory but it is restricted to the Talmud. Here their memory is prodigious. They have the ability to recall, from memory, every word on every page of the twelve volumes of the Babylonian Talmud.

There are two features that characterize memorists. First, although their memory may be good for some subject matter, they do not usually have a *deep* understanding of it. Second, having an exceptional memory is unpleasant. This was clear in Luria's subject Shereshevskii and is also well captured by the Argentinian writer Jorge Louis Borges in his short story "Funes, the Memorious." Here Borges describes a young man with a perfect memory that proved to be a tragic burden:

> We, in a glance, perceive three wine glasses on the table; Funes saw all the shoots, clusters, and grapes of the vine. He remembered

the shapes of the clouds in the south at dawn on the 30th of April in 1882, and he could compare them in his recollection with the marbled grain in the design of a leather-bound book which he had seen only once, and with the lines in the spray which an oar raised in the Rio Negro on the eve of the battle of the Quebracho. These recollections were not simple; each visual image was linked to muscular sensations, thermal sensations, etc. He could reconstruct all his dreams, all his fancies. Two or three times he had reconstructed an entire day. He told me: I have more memories in myself alone than all men have had since the world was a world . . . My memory, sir, is like a garbage disposal.

But exceptional memory is not limited to memorists. A more common type of good memory (called *flashbulb memory*) is something most people have at various times in their lives. A flashbulb memory is a detailed and vivid memory that is stored on one occasion and retained for a lifetime. Flashbulb memories, such as the memory of where you were and what you were doing when you heard that President Kennedy was assassinated or that the *Challenger* space shuttle had exploded, preserve knowledge of an event in a long-lasting way. Initial studies of flashbulb memories focused on important historical events. But there is also evidence that we can retain autobiographical information about surprising and defining personal events with the same vivid clarity. These memories are not necessarily accurate in every respect, but it is clear that the brain is able to enhance the power of long-term memories formed by emotionally significant events.

How are the details of dramatic personal and historical events stored? These surprising and emotionally charged events are thought to depend on the amygdala, a brain structure that we shall learn more about in Chapter 8, as well as the major arousal systems of the mammalian including the human brain—modulatory systems

that use the release of the neurotransmitters serotonin, norepinephrine, dopamine, and acetylcholine to regulate mood and alertness. One potential consequence of the action of these modulatory systems might be to relieve repression by CREB-2 and thereby prime the memory system so that a single experience is sufficient to put information into long-term memory. It is therefore of interest that these modulatory transmitter systems can play a significant role in the sort of learning that CREB is involved with in *Aplysia, Drosophila,* and as we shall see below, in mice.

GENES AND PROTEINS FOR LONG-TERM MEMORY

The switch for long-term memory is thrown only part way by removing the repressor (CREB-2) and activating the activator (CREB-1). To throw it all the way, the genes activated by CREB-1 must yield their protein products. It is this step—when the proteins encoded by the genes that are activated by CREB-1 are synthesized—that is thought to be the sensitive (consolidation) phase during the forming of long-term memory. Before this step is completed, memory can be blocked by inhibitors of protein synthesis, and a blow to the head can forever erase a memory in the process of being consolidated.

As we have seen, new protein synthesis is required for only a brief period, lasting one to several hours. The rapidity of this mechanism suggested to Goelet and Kandel that the proteins must be manufactured and needed for only a brief period. This idea led them to propose that the consolidation phase of memory is a period during which CREB-1 activates a special class of genes called *immediate response* genes, or *immediate early* genes. These genes are characterized by the fact that they are activated rapidly and transiently.

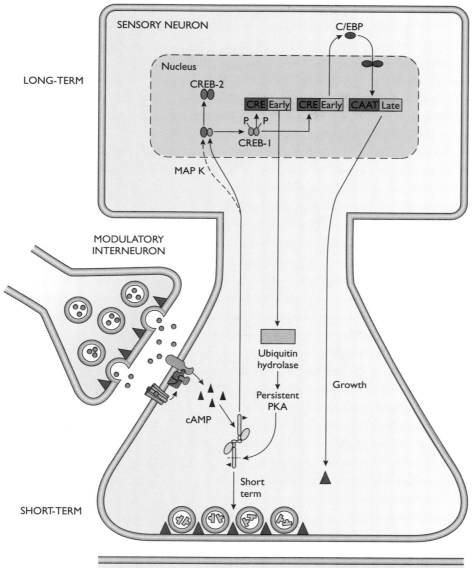

Long-term sensitization of the gill-withdrawal reflex of Aplysia *is initiated with repeated activation of the sensory neurons by serotonergic modulatory interneurons. This leads to two major sets of genetically induced changes in the sensory neurons of the reflex: (1) the persistent activity of protein kinase A and (2) the growth of new synaptic connections. In the short-term process (Chapter 3), serotonin released from modulatory interneurons initiates a chain of steps resulting in an increase in cAMP and temporary activation of PKA, a protein kinase that acts to enhance transmitter release. With repeated release of serotonin, PKA is also translocated to the nucleus, where it initiates the switch for the long-term process. There PKA is thought to lead to removal of the inhibitory action of the repressor CREB-2 by its ability to activate the MAP kinase. PKA also activates the CREB-1 protein, a transcription regulator that initiates transcription of various genes. A product of one of these genes, ubiquitin hydrolase, acts on the (inhibitory) regulatory subunit of PKA itself to keep PKA active. Another product of these genes is a transcription factor, C/EBP, that acts on genes farther downstream to initiate the growth of new synaptic connections.*

Christina Alberini, Kaoru Inokuchi, Ashok Hedge, James Schwartz, and Kandel therefore screened for and found two immediate-response genes in *Aplysia* that are rapidly induced at the time that long-term facilitation is established and that could be actived by cAMP and CREB-1. One gene encodes an enzyme, ubiquitin hydrolase, and the other a transcription regulator called C/EBP. Blocking the expression of either of these genes blocks the long-term process without affecting the short-term process. Study of these two genes has proven extremely informative.

The enzyme ubiquitin hydrolase, encoded by the first of the genes, sets up a positive feedback action. This enzyme is part of a protein complex, called the ubiquitin proteasome, that selectively destroys proteins. Schwartz and his colleagues had previously shown that in the sensory neurons, the ubiquitin proteasome destroys the regulatory subunits that normally inhibit the catalytic subunit of the PKA. As we saw, this kinase is critical for establishing long-term memory. By destroying the regulatory subunits, the ubiquitin proteasome thus removes a *second* inhibitory constraint on long-term memory (CREB-2 being the first).

You might have supposed that regulatory subunits had already been put out of action earlier, when they were bound by cAMP, when cAMP was at high levels. By the time ubiquitin hydrolase is activated, however, cAMP has begun to return to its normal level, and the regulatory subunits are starting to actively inhibit the action of PKA again. Worse, any protein phosphorylated by PKA has been rapidly dephosphorylated by an opposing enzyme, called a phosphatase, that removes phosphate groups from proteins. With the removal of the regulating subunit by the ubiquitin proteosome, however, the kinase is now able to be continuously active, even though the level of cAMP has returned to normal. As a result, the counteracting action of the phosphatase is overridden, and the kinase can continue to phosphorylate target proteins. Indeed, David Sweatt and Kandel found that many proteins that are phosphorylated in the short term can be maintained in a phosphorylated state in the long term, without the assistance of cAMP. We have here what is perhaps the simplest molecular positive-feedback mechanism for long-term memory. The persistent kinase is particularly important during the first several hours of long-term memory storage, when the growth of new synaptic connections is beginning.

The second immediate response gene activated by the long-term memory switch is the gene for the *Aplysia* transcription regulator C/EBP, a gene activator. Blocking the expression of the *Aplysia* gene for C/EBP blocks the growth of new synaptic connections. These two genes are probably only the tip of the iceberg. Many genes are likely to be expressed during the transition to long-term memory.

GROWING NEW SYNAPSES

Once long-term memory in *Aplysia* is induced, it can be maintained stably for days and weeks, depending on the extent of training. What makes long-term memory storage so stable?

Looking at sensitization of the gill withdrawal reflex, Craig Bailey and Mary Chen demonstrated that what maintains this memory stably in *Aplysia* are alterations in cell anatomy. Following a training procedure that produces memory for sensitization lasting three weeks, there is a doubling in the number of synaptic terminals per sensory neuron, from 1300 to 2600. These changes persist for as long as the change in the gill withdrawal reflex lasts. As the memory decays over the three-week period, synaptic terminals are lost, and they gradually regress back to their initial number. These anatomical changes are *not* limited to the presynaptic sensory neurons. In sensitized animals, the dendritic processes of the postsynaptic cell also grow to accommodate the new synaptic growth. Thus, a characteristic feature of long-term memory stor-

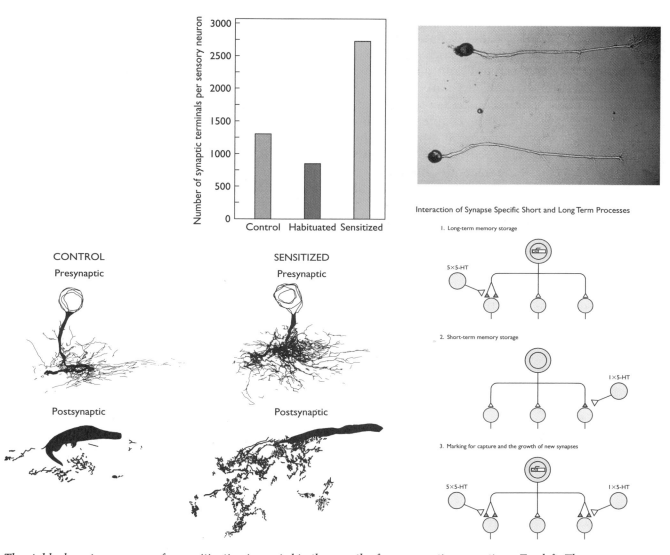

The stable, long-term memory for sensitization is carried in the growth of new synaptic connections. Top left: The sensory neurons of sensitized animals have many more presynaptic terminals than control or habituated animals have. Bottom left: Both sensory neurons (top) and motor neurons (bottom) grow new branches in the course of sensitization. Top right: The photomicrograph shows the culture system developed in Aplysia that allows one to examine the actions of two independent branches of a single sensory neuron (small neuron in the middle of the photomicrograph) on two different motor neurons (large neurons at the top and bottom of the photomicrograph). Serotonin can be selectively applied to only one and not the other of the two branches of the sensory neuron. Bottom right: The three drawings illustrate the consequences of activating the short-term process by one pulse of serotonin (1 × 5-HT), or the long-term process by five pulses of serotonin (5 × 5-HT), or of activating both in conjuction with one another.

age is that it requires a coordinated structural change in both pre- and postsynaptic cells.

It is easy to see that by having twice as many synaptic connections, any given sensory neuron will be twice as effective in exciting gill motor neurons. As a result, a touch to the siphon will be much more likely to produce a gill withdrawal reflex, and the enhancement will persist as long as the additional synaptic connections persist. Interestingly, long-term memory is not always reflected in the growth of additional synapses. In the case of long-term habituation, described in Chapter 2, there is a pruning of synaptic terminals from the original 1300 per neuron to a mere 800 per neuron.

As we shall see later in this chapter, structural changes may also be characteristic of the stable phase of declarative memory. These anatomical alterations reveal a fundamental difference between short-term and long-term memory. The alterations for short-term memory are limited to small subcellular changes, for example, shifts in the location of synaptic vesicles to place them closer to or farther from the active zone. Such shifts are thought to alter the cell's ability to release transmitter. In contrast, long-term memory is associated with the growth of new synaptic connections or the retraction of pre-existing ones. Thus, at the cellular level, the switch from short-term to long-term facilitation is a switch from a *process-based* memory to a *structure-based* memory.

A SYNAPSE-SPECIFIC MECHANISM FOR LONG-TERM MEMORY

The nucleus is a resource shared by all the synapses within the cell. The requirement that genes in the nucleus be transcribed to establish long-lasting memories has posed a large and important question in the study of memory. Do the synaptic changes contributing to long-lasting memory need to be cellwide, or can the strength of individual synaptic connections be modified independently? Each of the approximately 10^{11} neurons in the brain makes an average 1000 synaptic connections on its population of target cells. It has been thought that, to maximize information processing, the critical unit of synaptic

plasticity would be the individual synapse. Is that the case, and, if so, how are the genes in the nucleus able to target a single synapse?

To address this question, Kelsey Martin, Casadio, Bailey, and Kandel cultured a single *Aplysia* sensory neuron that bifurcated and made synaptic contact with two spatially separate motor neurons. By perfusing five pulses of serotonin onto the synapses made onto one of the motor neurons, they found that the single axonal branch at that motor neuron undergoes long-term facilitation, while the other axonal branch does not change. New synaptic connections grow exclusively at the treated branch. It appears that a signal goes back to the nucleus from this synaptic region. The signal then activates gene products that go to all the terminals, but only those terminals that have undergone a recent short-term change can use the protein coming down from the nucleus to those synapses for the purpose of making long-term structural changes.

To test this idea, Martin, Casadio and their colleagues next applied five pulses of serotonin to one set of contacts made by the sensory neuron onto one motor neuron and then applied a single pulse of serotonin, which by itself produces only transient, short-term synaptic facilitation lasting minutes, to the contacts made by the sensory neuron onto the second motor neuron. In this context, the single pulse of serotonin was able to recruit and capture, for the second branch, long-term facilitation and the growth of new connections.

A SWITCH FOR LONG-TERM DECLARATIVE MEMORY

A trained *Aplysia* vigorously withdraws its gill, and a trained *Drosophila* successfully avoids an inappropriate odor. In both cases, similar molecular mechanisms have created a long-term nonde-

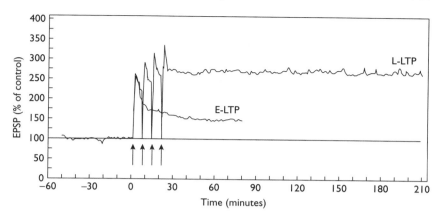

Using the same experimental setup shown on page 112, investigators have recorded both an early and a late phase of LTP in the Schaffer collateral pathway leading to the CA1 region of the hippocampus. A single train of stimuli given for one second at 100 Hz elicits early LTP (E–LTP), and four trains given at 10-minute intervals elicit the late phase of LTP (L–LTP). The resulting early LTP lasts two hours, and late LTP more than 24 hours.

clarative memory. In each case, a second-messenger, the cAMP-dependent protein kinase, has translocated to the nucleus of specific neurons, where it activates genes that trigger the formation of new synaptic connections. Let us now turn our attention to declarative forms of memory. When a student memorizes a poem or a mouse learns to navigate a maze, do similar mechanisms take place in neurons? Have these mechanisms been conserved during evolution as higher forms of memory evolved from lower ones?

As we discussed in Chapter 6, each of the major pathways of the hippocampus, a central component of the medial temporal lobe system, is capable of long-term potentiation (LTP)—a synaptic mechanism that seems suitable for participating in declarative memory storage. LTP is a persistent, activity-dependent form of synaptic modification that can be induced by brief, high-frequency stimulation of hippocampal neurons. We saw in Chapter 6 that genetic interference with the initial stages of LTP disrupts aspects of both short-term as well as long-term memory: a mouse does not learn to find its way around a maze, much as the gill withdrawal reflex of an *Aplysia* does not become sensitized if short-term facilitation is blocked at its very onset. Is there a distinct long-term component of LTP, analogous to long-term facilitation in *Aplysia?*

With this question in mind, Uwe Frey, Yan-You Huang, and Kandel examined LTP in the Schaffer collateral pathway of rat hippocampal slices and found distinct temporal phases, much like short-term and long-term facilitation in *Aplysia.* There is an early phase beginning immediately after tetanic stimulation and lasting one to three hours. This phase is induced by a single high-frequency train and does not require protein synthesis. By contrast, three or more high-frequency trains induce a late phase (L-LTP) that persists for at least twenty-eight hours, and this late phase shows all the signs of requiring gene activation: it is blocked by inhibitors of protein synthesis, by inhibitors of RNA synthesis, and by inhibitors of PKA. Conversely, this late phase can be activated by cAMP, one of the messengers in the second-messenger system that sends a signal to the nucleus to begin activating genes.

As is the case with long-term facilitation in *Aplysia,* the late phase of LTP in the rat hippocampus includes some initial steps during which new protein is synthesized. No matter whether one tries to induce L-LTP by tetanic stimulation or by applying cAMP, its induction is prevented when the transcription of genes is blocked, either immediately after tetanic stimulation or during the application of cAMP. Thus, the late phase of LTP requires gene transcription during a

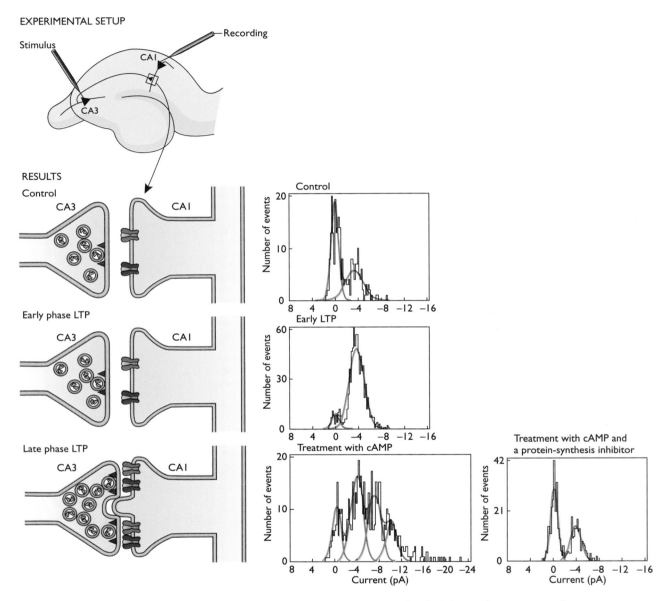

A distinction between the early and late phases of LTP is evident at the level of the single connections between a CA3 cell and a CA1 cell. As illustrated on page 115, a single CA3 cell is stimulated to produce a single elementary synaptic potential in a CA1 cell (upper left). When the CA3 cell is stimulated repeatedly at low frequency, it gives rise either to an elementary synaptic potential (the release of a single quantum), measured by the electric current in the CA1 cell as about 4 picoamperes (pA), or to a failure (no release), measured as 0 pA. In control cells, there are many failures, as indicated in the graph on the top by the high curve centered around the 0 point. The distribution of failures and successes suggests that the synapse has a low probability of releasing vesicles, as indicated in the drawing at the top left (control). As we have seen on page 115, in the early phase of LTP, the distribution of responses suggests that a single release site now releases a vesicle with a high probability. This is indicated in the middle drawing on the left (early phase LTP). When the late phase of LTP is induced by cAMP, the distribution of responses suggests the possibility that new presynaptic active zones and postsynaptic receptors have grown. This is indicated at the bottom left (late phase LTP). These late phase effects are blocked by an inhibitor of protein synthesis.

critical period immediately following tetanization, perhaps because special genes must be expressed during this period. To test this possibility, Joe Tsien, Dietmar Kuhl, and Kandel screened mice for immediate response genes induced by LTP and found that a number of such genes were induced. Two are particularly important, and both have a CRE element in the regulatory region of the gene. One of these genes encodes tissue plasminogen activator (tPA), an enzyme that has been found to stimulate the growth of axon terminals and dendritic spines, and the other is a gene related to C/EBP of *Aplysia*, a gene that we have already seen is critical for the switching on of long-term facilitation in that animal.

This late phase of LTP was first delineated by recording the synaptic response of many cells simultaneously. How is this novel phase manifest in the elementary synaptic connections between individual cells? As we saw in Chapter 6, a single presynaptic CA3 neuron in the unstimulated, normal state makes a single synaptic connection with its target cell in the CA1 region, and this one synaptic terminal seems to release only a single vesicle of transmitter, from a single release site, in an all-or-none fashion. In the normal state, there are many failures of release and few successes. After the induction of the early phase of LTP, there are far fewer failures of release and many more successes. The most straightforward interpretation of this result is that the early phase of LTP results from an increase in the probability of vesicle release without the addition of new release sites.

Does the late phase of LTP, examined several hours after induction, also involve a persistent increase in the probability of vesicle release? Vadim Bolshakov, Steven Siegelbaum, Hava Golan, and Kandel examined the consequences of treating a hippocampus slice with cAMP and inducing the late phase in essentially all synapses. They found that the fraction of successful releases was consistently higher than before treatment; these results suggest that at some synapses there continues to be a high probability of transmitter release during the late phase of LTP. In addition, they were surprised to find that the synaptic responses could no longer be described by a single (Gaussian) curve distribution reflecting a single release site. One likely interpretation is that the late phase of LTP requires a growth of new synaptic release sites: the addition into the presynaptic terminals of new release sites and the insertion of new receptors into the dendritic spines of the postsynaptic cell.

STRUCTURAL CHANGES IN THE LATE PHASE OF LTP

These findings suggest the interesting possibility that, with the addition of new release and receptive sites in the late phase of LTP, there is an increase in the number of synaptic contacts. It has been suggested that such an increase could occur through the splitting of the single, pre-existing active zone into two by the growth of the postsynaptic spine. In fact, Yuri Geinisman at Northwestern University Medical School and his colleagues have observed an increase in the number of exactly this sort of synapse after inducing LTP. At these synapses, a projection or spinule from the postsynaptic spine juts into the presynaptic terminal, splitting the active zone into two discrete regions. It seems that one particular class of synapse has been transformed into another.

The tendency to grow new anatomical connections in response to experience turns out to be a ubiquitous feature of mammalian brains. As early as 1990, three different groups—headed by Gary Lynch, William Greenough, and Per Andersen—had provided evidence that the growth of new synapses in the hippocampus is correlated with LTP. As we shall learn in Chapter 10, structural changes are the signature of long-term memory for many types of experiences.

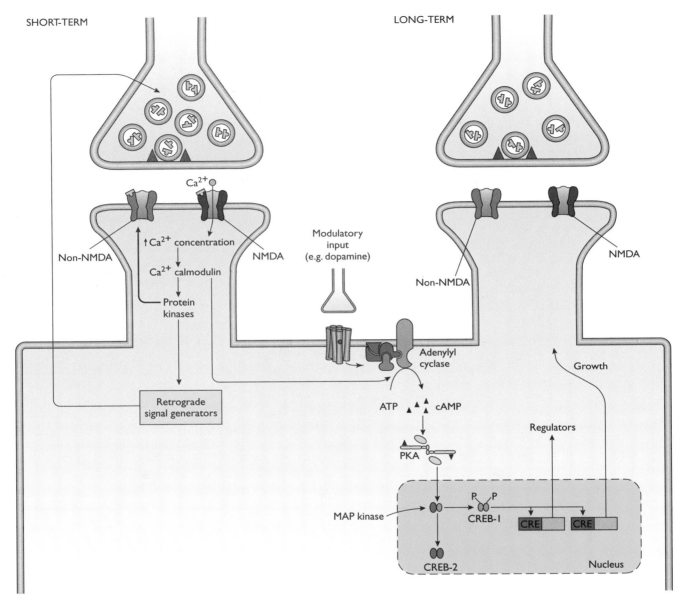

A model for the early and late phases of LTP. A single train of stimulation leads to the activation of early LTP by activating NMDA receptors and Ca^{2+} influx into the postsynaptic cell. The Ca^{2+} combines with calmodulin to activate a set of second messenger protein kinases that are thought to mediate at least two functions. One, the kinases phosphorylate AMPA receptor channels and thereby increase the sensitivity of these post-synaptic receptors to glutamate. In addition, the kinases are thought to activate a set of enzymes that generate retrograde signals that feed back on the terminals of the presynaptic neuron to enhance transmitter release. With repeated trains, the Ca^{2+} influx also engages adenylyl cyclase. That enzyme activates PKA, which then translocates to the nucleus where it is thought to phosphorylate CREB. CREB in turn activates targets that are both regulators and effectors of growth and are thought to lead to structural changes.

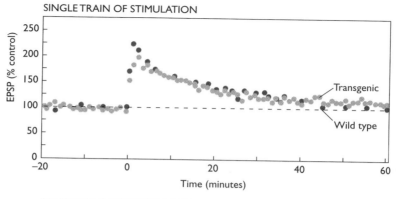

SINGLE TRAIN OF STIMULATION

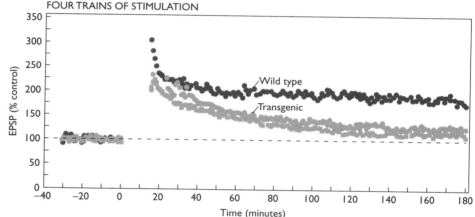

FOUR TRAINS OF STIMULATION

The early phase of LTP is normal both in wild-type mice and in transgenic mice that express a gene blocking the action of PKA (top graph). The reduction in PKA activity eliminates the late phase of LTP in transgenic mice (bottom graph).

A MODEL FOR THE LATE PHASE OF LTP

Thus, in the late phase of LTP, both the presynaptic and postsynaptic elements of the synapse seem to undergo a coordinated long-term change. These studies suggest the outlines of a molecular model according to which LTP in the hippocampus has phases much like facilitation in *Aplysia*. The early phase of LTP involves the activation in the postsynaptic cell of several protein kinases unrelated to PKA. This *postsynaptic* initiation in turn is thought to lead to an insertion of new AMPA receptors and thereby to an increase in the sensitivity of postsynaptic receptors to glutamate as well as to an increase in the amount of transmitter released, perhaps through the action of

one or more retrograde messengers, which communicate back to the presynaptic cell from the postsynaptic cell.

With repeated trains of stimulation in the pathways of the hippocampus, however, something new starts to happen. As the late phase of LTP kicks in, levels of cAMP shoot up, as in *Aplysia* and *Drosophila,* and this increase in cAMP in the hippocampus is followed by activation of PKA and CREB-1. Finally, much as in *Aplysia,* the activity of CREB-1 in the hippocampus appears to lead to the activation of a set of immediate response genes, and these genes act to initiate the growth of new synaptic sites.

So far we have relied on pharmacological agents and biophysical studies to distinguish the late phase of LTP from the early phase and to

TRAINING

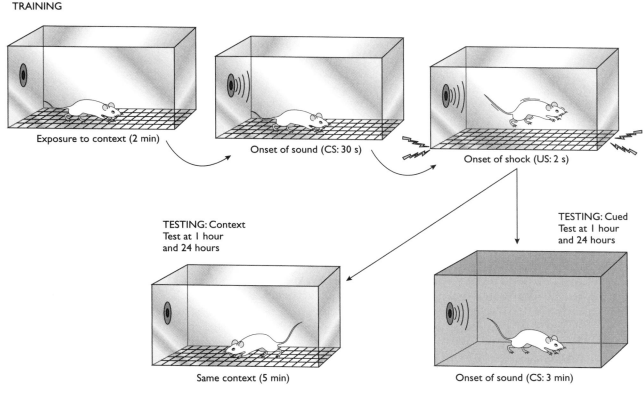

The protocol for two types of fear conditioning that can be produced following a single learning trial. The trial consists of a single foot shock and produces context conditioning (to a box) and cued conditioning (to a tone).

analyze the mechanisms of the late phase. However, these analyses are potentially limited in two ways. First, pharmacological agents are rarely completely specific. They sometimes affect targets other than those against which they are intended. Genetic experiments are in this regard much more precise. Moreover, the pharmacological and biophysical analyses we have so far considered have not yet revealed what relation, if any, this phase of LTP has to long-term memory.

As we indicated in the last chapter, the ability to generate genetically modified mice has proven to be important. We can selectively manipulate individual genes in an intact animal and see how that affects the animal's behavior—its ability to

store long-term memory. Two sets of studies with this goal have been carried out in genetically modified animals. Both studies set out to eliminate the late phase of LTP without blocking the early phase. Ted Abel, Mark Barad, Rusiko Bourtchouladze, and Kandel generated mutant mice expressing a gene that blocked the action of the catalytic subunit of PKA. In addition, Bourtchouladze and Alcino Silva at Cold Spring Harbor studied mice that had been altered by a partial knockout of CREB-1. Both sets of transgenic animals had roughly similar defects in LTP. In area CA1, the early phase was normal, but the late phase, instead of lasting for many hours as it does normally, decayed back to baseline within one or two hours. These findings provided independent evidence that the late phase of LTP depends on PKA and on specific gene actions initiated by CREB-1.

We now have, for the first time, animals with a normal early phase of LTP and some defect in the late phase. What about their memory capabilities? We might predict that they could learn well and have good short-term memory, but then exhibit a defect in some type of long-term memory.

In fact, this is exactly the case. These mice had a serious defect in long-term spatial memory. However, the typical maze tasks are not ideal for testing short-term memory since they require repeated training over several days. Thus, they do not provide the temporal resolution necessary to identify clearly the early phase of memory storage. Bourtchouladze and Abel therefore turned to conditioning tasks that trigger robust learning in a single trial. Animals trained in these tasks learn to fear a new environment (a context) as well as to fear a neutral conditioned stimulus (CS), such as a tone, because the context and the tone are coincident in time with an aversive unconditioned stimulus (US), usually foot shock. Specifically, a mouse is placed in a small box with a grid floor that can be electrified to deliver a mild foot shock. The mouse explores for two minutes, becoming familiar with this new environment. A tone is then presented, followed by a shock to the feet. When later placed in the same box, animals exhibit fear by crouching and becoming immobile—they freeze. This form of declarative, contextual memory requires the hippocampus. Similarly, the animal also learns to fear the tone and will freeze in any context if the tone is heard. This form of nondeclarative fear conditioning requires the amygdala.

In these tasks, both types of genetically modified mice learned to freeze as easily as normal mice, and an hour after training they still froze at the sound of the tone or the sight of the box—their short-term memory was normal. But 24 hours later they no longer reacted to the box—

they were deficient in long-term memory for context, which requires the hippocampus. By contrast, they responded normally to the tone that depends only on the amygdala, a region where the transgene was not expressed. Because one set of genetically modified mice expresses a gene that inhibits PKA, the experiment shows that this protein kinase is critically important for the conversion of short-term memory into long-term memory, perhaps because the kinase phosphorylates transcription factors like CREB-1 that in turn activate proteins required for long-lasting LTP. This idea is supported by the findings from the second set of mice studied by Bourtchouladze and Silva. This set of mice had impaired memory secondary to a partial knockout of CREB-1, thereby also suggesting that one of the genes turned on by PKA is CREB-1.

Abel, Barad, Bourtchouladze, and Kandel asked if the time window for long-term memory formation in these mice is the same as the time window for the late phase of LTP that is dependent on protein synthesis. They found that mice learned the freezing response initially and retained the response an hour after training despite having been injected with a protein synthesis inhibitor before training. By contrast, when they were tested 24 hours after training, the same mice showed little fear response, and thus showed a dramatic deficit in memory. The inhibitors of protein synthesis were effective when given during or immediately after training, not when given one hour following training. The implication is that the time course of residual fear memory that is evident when inhibitors of protein synthesis are given during or shortly after training parallels the time course of residual fear memory observed in the transgenic animals. This suggests that PKA is essential for the formation of long-term memory.

Work by James McGaugh at the University of California, Irvine, and by Yadin Dudai at the Weitzmann Institute have importantly extended these conclusions. They have found that emotional memory involving the amygdala and a type

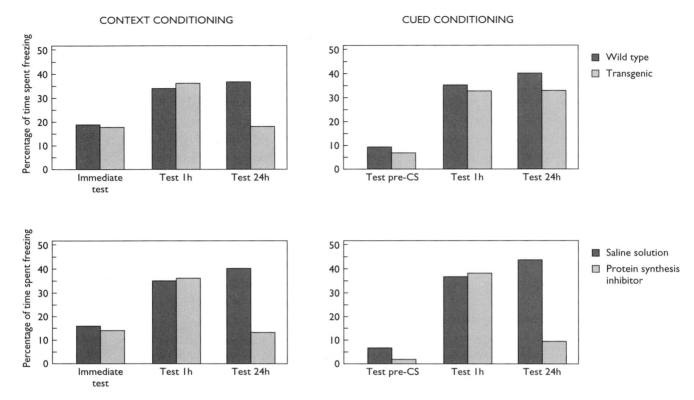

Mutant mice that express a gene in the hippocampus blocking the action of PKA, or mice that have been given a protein synthesis inhibitor, have been conditioned to freeze both to context in the form of the sight of a box (left) and also to a cue given as a tone (right). These mice have a good short-term memory for fear conditioning at one hour. However, they no longer react to the box 24 hours after conditioning, indicating a long-term memory defect in a form of declarative memory that requires the hippocampus (left top). Yet 24 hours after training, these same mice still freeze in response to the tone because the tone conditioning represents a form of nondeclarative conditioning (right top) that requires the amygdala, a structure in which the transgene is not expressed. By contrast, inhibitors of protein synthesis block long-term memory for both cued and context conditioning because the inhibitor affects the hippocampus and the amygdala similarly.

of memory for avoiding poisoned food called bait shyness that involves the taste area of the cerebral cortex both require the CREB-mediated switch for long-term memory.

A CONSERVED SET OF MOLECULAR MECHANISMS

We have seen that declarative and nondeclarative forms of memory are expressed very differently in terms of behavior. Moreover, these two forms of memory each use a different cognitive logic (conscious recollection on the one hand and unconscious performance on the other) and different anatomical systems. Yet, despite all the differences, there is a surprising degree of similarity not only in their elementary storage mechanisms, but, specifically, in the switch they use to convert short-term to long-term memory.

To begin with, both declarative and nondeclarative memory go through phases in memory storage that are clearly reflected in an animal's behavior. There is a labile, short-term phase and a stable, self-maintained, long-term phase. Second, repetition helps to convert the short-term to the long-term phase. Third, with both types of storage these two phases are evident at individual synapses, and in each case the duration of these synaptic changes roughly mirrors the lifetime of the two phases of memory storage.

There are also similarities with regard to molecular mechanisms. With both forms of memory, short-term memory is achieved by modifying pre-existing proteins and strengthening pre-existing connections through the activity of one or another protein kinase. These short-term forms of memory do not require new protein synthesis. By contrast, long-term memory requires the activation of genes, new protein synthesis, and the growth of new synaptic connections.

The capability for long-term change arises in each case because the synapse has privileged communication with the cell nucleus and the machinery for gene expression. Indeed, both nondeclarative and declarative memory seem to use a common molecular signaling cascade for communication from the synapse to the nucleus. The participants in this cascade include at least one second-messenger cAMP, two protein kinases (PKA and MAP kinase), and the transcription activator CREB-1. In each case, CREB-1 acts to induce immediate response genes that encode proteins essential for the growth of new synaptic connections.

These several findings have provided a new set of insights into the evolutionary conservatism underlying the molecular underpinnings of mental processes. The simplest memory capabilities, and those that seem to have appeared earliest in evolution, seem to be nondeclarative memories related to survival, feeding, mating, defense, and escape. As a variety of additional types of nondeclarative memory and then declarative memory evolved, the new memory processes retained not simply a set of genes and proteins, but entire signaling pathways and programs for switching on and stabilizing synaptic connections. Moreover, these common mechanisms have also been conserved through the evolutionary history of species; they are found in both simple invertebrates such as *Drosophila* and *Aplysia* and complex mammals such as mice.

THE TWO SIDES TO THE BIOLOGICAL STUDY OF MEMORY

As we outlined in the first chapter, the biological analysis of memory has two parts: (1) there are the elementary *molecular mechanisms of memory storage* whereby cells and synapses are modified in the long term in response to learning; and (2) there are the *systems of memory storage* that determine how these elementary mechanisms are recruited and used in storing and retrieving declarative and nondeclarative information.

As we have seen in this and in earlier chapters, we are now beginning to have some insights into a number of the elementary mechanisms of storage for both declarative and nondeclarative memory. What about the systems properties of memory? Here we need to know: What structures and connections are important for declarative and nondeclarative memory? Where in the systems is memory stored?

It is to these questions that we now turn. We have previously considered simple invertebrate systems for nondeclarative memory in Chapters 2 and 3 and brain systems for declarative memory in Chapter 5. We will see in the next two chapters that there also are several types of nondeclarative memory, each calling on its own brain system.

Henry Matisse, Memory of Oceania (1953). Matisse (1869–1954), renowned for his revolutionary use of color to evoke emotion, here recollects a pleasurable visit to Tahiti, reminding us of the emotional aspects of memory.

8 | Priming, Perceptual Learning, and Emotional Learning

As we learned in Chapters 1 and 2, one of the fascinating features of information processing in the brain is that much of what is processed is not accessible to conscious awareness. Consider, for example, the process whereby an object such as a pencil is identified, located, and reached for. In Chapter 5, we learned that more than one brain system supports these visual abilities. A ventral stream of processing reaches the temporal lobe and is concerned with visual identification of the pencil. A dorsal stream of processing reaches the parietal cortex and is concerned with locating the object in space and with using vision to navigate through space. Though we experience perception as a unified conscious phenomenon, our perceptions are constructed from several component parts.

Only some of these components are accessible to conscious experience. Conscious visual experience appears to be associated with processing in the *ventral* stream. That is, we become aware of the shape and color of a pencil as a result of the same neural operations by which we come to recognize and identify pencils in our visual world. In contrast, some of the knowledge we achieve from the *dorsal* stream of processing, in the form of the motor programs needed to reach for a pencil, is not accessible to awareness. The knowledge we obtain about how and where to reach for the pencil is unconscious.

This idea is illustrated by a simple study of reaching carried out by Melvyn Goodale at the University of Western Ontario, Canada. Volunteers viewed two displays, each presenting a plastic disk surrounded by a circular array of smaller or larger disks. When arranged in this way, the disks create a familiar optical illusion that was first described by Hermann Ebbinghaus. When a disk is surrounded by larger disks, it appears smaller than its actual size, and when the same disk is surrounded by smaller disks it appears larger than its actual size.

In the study, subjects were asked to pick up the central disk from the right display when the two central disks appeared to be the same size and to pick up the central disk from the left display when the two central disks appeared to be of different size. Which display was on the left and which was on the right varied randomly from trial to trial. To pick up the disks, the subjects used their thumb and forefinger, and the size of the finger–thumb gap (the grip aperture) was recorded on videotape.

The results were clear. The choice of which disk to pick up (the one in the left display or the one in the right display) was consistently under the control of the illusion. Across many trials of testing, subjects treated disks that were actually physically identical as physically different, and they treated disks that were physically different as physically identical. The way that subjects reached told a very different story. Specifically, the size of the finger–thumb gap as they reached for

In the standard version of the circles illusion, shown at the top, the disks in the centers of the two arrays are physically identical even though they appear to be different in size. In the lower display, the central disk in the array of larger circles on the right is actually a little larger than the central disk to the left. However, most observers rate the two central disks as perceptually equivalent in size. Nevertheless, when subjects reach to pick up one of the central disks in the lower display, they make their thumb–finger grip smaller or larger in accordance with the actual size of the disk.

the disks was immune to the optical illusion. Grip size was consistently determined by the true size of the target. For example, on those trials when subjects perceived the two central disks as being the same size, but one disk was actually 2.5 millimeters larger in diameter than the other, grip size followed the true size of the disks: subjects opened their grip wider when reaching toward a physically larger disk than when reaching toward a physically smaller one.

These studies indicate that while conscious visual perception is associated with recognition and identification, and with the function of the ventral stream of visual processing, other aspects of visual information processing are unconscious and do not afford awareness. These insights about the nature of visual perception apply as well to the faculty of memory. Memory is not a single entity but is composed of different systems. Only one of these systems is accessible to awareness, the declarative memory system. Chapter 1 introduced the idea that some forms of memory, such as motor skill learning, are not accessible to awareness. Chapters 2 and 3 showed how the cellular processes underlying some simple forms of unconscious memory (habituation, sensitization, and classical conditioning) can be studied in invertebrate animals with relatively simple nervous systems. This chapter considers three additional kinds of nondeclarative memory, which are exhibited by humans and other vertebrate animals: priming, perceptual learning, and emotional learning. Whereas invertebrate animals seem to have available only nondeclarative memory, what makes the study of memory in humans and other higher vertebrates particularly fascinating is that they have developed a strong capacity for declarative memory (the subject of Chapters 4, 5, and 6) but retain the capacity for nondeclarative memory as well.

Historically, it took some time for the idea that there might be nondeclarative, unconscious forms of memory to come into the foreground. In 1962 Brenda Milner found that the amnesic patient H.M. could learn the perceptuo-motor skill of drawing the outline of a star that was reflected in a mirror, but scientists tended to set aside these motor skill findings and continued to view the rest of memory as a single, undifferentiated entity. Demonstrations of unexpectedly good learning and retention by amnesic patients on other than motor skill tasks were reported in the late 1960s and in the 1970s, but for two reasons

these findings did not lead to proposals about new kinds of memory. First, even when the performance of amnesic patients was good, it often fell short of normal levels. In these instances, good performance could be explained by supposing that memory tasks vary in how sensitive they are to detecting what subjects have learned. Amnesic patients might sometimes perform well simply because some tests are good at detecting their residual learning and memory ability.

Second, it was unclear what test procedures were needed to obtain good performance in amnesic patients. In the 1970s, Elizabeth Warrington at Queens Square Hospital in London and Lawrence Weiskrantz at Oxford University found that amnesic patients sometimes performed normally on what we now call priming tasks—specially constructed memory tests that used three-letter word stems as cues (MOT, INC) to recover previously presented words (MOTEL, INCOME). This was another early demonstration of a separate form of memory. However, it took some time to appreciate the critical importance of the instructions given to subjects.

In 1980, Neal Cohen and Squire showed that amnesic patients could learn and retain as well as normal subjects the skill of reading mirror-reversed print. By moving beyond motor skills, this finding broadened the scope of what amnesic patients could do and suggested a major distinction between declarative forms of memory, which are impaired in amnesia, and nondeclarative forms of memory, which are preserved in amnesia. The implication was that some yet-to-be-explored collection of learning and memory abilities was independent of the medial temporal lobe brain structures that are damaged in amnesia. This chapter and Chapter 9 focus on the kinds of nondeclarative memory that have been discovered in human beings and how these kinds of memory are organized in the brain.

In the early 1980s, Endel Tulving and Daniel Schacter at the University of Toronto took up the problem posed by Warrington and Weiskrantz of why some kinds of memory tests for words appear to yield different results from others. They showed that the ability to recover recently studied words from letter fragments (for example, after studying the word ASSASSIN, it becomes easier to form a word from _SS_SS_) and the ability to recognize recently studied words as familiar could be separated in normal subjects. Soon thereafter, Peter Graf, George Mandler, and Squire used word lists and three-letter word stems to show how important the instructions are in tests of this kind. Only one kind of instruction results in normal performance by amnesic patients ("use this three-letter word stem to form the first word that comes to mind"). With conventional memory instructions ("use this three-letter cue to remember a recently presented word"), amnesic patients performed more poorly than normal subjects. Tests like the former, which amnesic patients consistently performed as successfully as normal subjects, came to be called tests of priming. By the time this study was done, in 1984, the idea that there may be many kinds of memory was in the air, and the phenomenon of priming quickly became a major topic of memory research.

PRIMING

Priming refers to an improvement in the ability to detect or identify words or objects after recent experience with them. By this definition, priming might appear to be just another way of speaking about everyday, declarative memory. However, careful studies have shown that priming is a distinct memory phenomenon. Priming's key feature is that it is unconscious. Its function is to improve the perception of recently encountered stimuli, but we need not be aware that the speed or efficiency of perception is improved.

David Mitchell and Alan Brown at Southern Methodist University showed college students drawings of common objects (for example, airplane, hammer, or dog) and asked them to name each object as quickly as possible. Later, they presented the same drawings again, mixed together with new drawings. The students took an average of about 0.9 second to name the new drawings, but they named the previously presented drawings in about 0.8 second, that is, almost one-tenth of a second faster than they named the new ones. There was no relationship between the ability to name a picture more quickly and the ability to recognize it as familiar. The priming effect is largely perceptual in nature. When two different examples of an airplane were presented, first a biplane and then a jet, the improvement in naming time was substantially reduced even though most people called both drawings an "airplane." Thus, the priming effect appears to depend mainly on the observer having carried out exactly the same perceptual operation on some earlier occasion. If an individual has processed a drawing of a dog, then it becomes easier to process exactly the same drawing of the dog when it appears a second time.

A remarkable feature of priming is that it can persist for an exceedingly long time even after a single experience. Carolyn Cave found that after naming 130 different pictures on one occasion, college students named those previously presented pictures more quickly than new pictures even a year later. The students showed improvement in naming times whether or not they could identify the pictures correctly as ones that had been encountered before.

If priming is independent of the ability to consciously remember, priming should involve brain systems other than the medial temporal lobe system that is essential for declarative memory. And amnesic patients, who are impaired at declarative remembering, should nevertheless be capable of long-lasting priming. Cave and Squire presented drawings of objects to amnesic patients

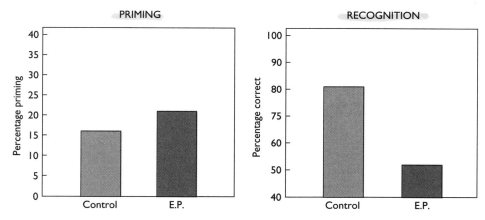

Priming and recognition in amnesic patient E.P. Left: Patient E.P. performed a little better than the average normal subject on 12 tests of priming. Seven normal subjects were tested. The priming score is the percentage of recently studied words that could be correctly read minus the percentage of nonstudied words that could be read. Right: When asked to identify the study words themselves in six separate tests of recognition memory, E.P. scored no better than he would have by guessing (50 percent correct).

on a single occasion and then one week later tested their naming speed for 50 previously viewed drawings and 50 new drawings. The patients named the 50 old pictures nearly 150 milliseconds faster than the 50 new pictures, despite the fact that the amnesic patients were very poor at recognizing which pictures had been presented before and which pictures were new.

Patients who are profoundly amnesic provide an opportunity to test just how different priming might be from declarative memory. Stephan Hamann and Squire gave tests of word priming to patient E.P. (see Chapter 1), and also gave tests of his ability to recognize words that had been presented. In each test session, 24 common English words were first presented for study and then either priming or recognition was tested five minutes later. In one priming test, 48 words were flashed briefly one at a time on a computer screen for about 25 milliseconds (24 old words and 24 new words). E.P. was able to read about 55 percent of flashed words that

had just been presented on the study list, but only about 33 percent of the words that had not appeared on the study list. Thus, having seen the words recently made them easier to read. Nonamnesic subjects performed identically. The findings for recognition memory were very different. In the recognition test, E.P. was shown 48 words one at a time and asked to say "yes" if the word had appeared on the study list and "no" if it had not. A score of 50 percent would represent chance performance, the score that would be obtained if one simply flipped a coin to determine each answer. Patient E.P. scored 52 percent correct on this test. Nonamnesicsubjects found this test quite easy and recog-nized about 80 percent of the words from the study list. Overall, E.P. was unable to recognize the words from the study list, but he showed fully normal priming for these same words.

The question inevitably arises why E.P., or other amnesic patients, cannot make use of priming to assist them in recognizing which items are familiar. Clearly, they have a record in their brain of their recent experience with words, because otherwise they would not be able to read the familiar words faster. When they are asked to judge the familiarity of a previously encountered word, why aren't they able to sense their fluency with the word and then judge that word familiar? The answer appears to be that individuals do not, perhaps cannot, consult the system that supports priming when making recognition judgments. When subjects are judging the familiarity (or pastness) of words, they are in a different mode of operation than when they are reading words. Moreover, how easily someone reads a word is actually not a reliable indicator of whether that word has been encountered recently. Many factors determine how quickly a word can be read, including its frequency in the language, how much the perceiver likes the word, and his or level of alertness and motivation. From this perspective, it is perhaps desirable that priming should not have much influence on declarative memory. Priming is presumably advantageous because animals evolved in a world where stimuli that are encountered once are likely to be encountered again. Priming improves the speed and efficiency with which organisms interact with a familiar environment.

The examples described so far have involved priming of drawings and words. In fact, priming can be demonstrated for almost any perceptible stimulus: pronounceable nonwords, strings of nonsense letters, unfamiliar visual objects, novel line patterns, and material presented by voice. In addition, priming can be observed on tests that require the analysis of meaning. For example, when asked to free associate to the word "child," people will say the word "baby" about 22 percent of the time. However, if the word "baby" has just been presented within a word list, the probability of saying "baby" when free associating to "child" is twice as great (45 percent). Amnesic patients have this capability at full strength, even though they fail to recall the word "baby" on a conventional memory test.

If priming reflects a separate form of memory, and one that is independent of declarative memory, where in the brain does priming occur? This question has been addressed by exploring the phenomenon of word stem completion priming, using the imaging technique of positron emission tomography (PET). Subjects first study words and are then given three-letter word stems with instructions to complete each stem with the first word that comes to mind. Subjects show the effects of priming in that they tend to complete the stems with words they have recently studied. Squire, Marc Raichle, and Raichle's collaborators at the Washington University PET facility obtained images of the brain while subjects were completing word stems. In one condition (baseline), none of the stems could form words from the previously presented study list. In another condition (priming), many of the stems could form words from the study list.

When PET images taken during the priming condition were compared to those taken during the baseline condition, there was a noticeable reduction in activity in the visual cortex at the back of the brain, in a region called the lingual gyrus. This finding was consistent with the idea that priming can be highly visual and that it occurs early in the visual processing pathways before the analysis of meaning. One simple way of thinking about priming is that for a period of time after a word or other perceptual object is presented, less neural activity is required to process that same word or object.

The PET technique provides information about *where* in the brain activity related to priming occurs, but it does not provide good information about *when* the activity occurs. Accordingly, it is conceivable that the reduced activity seen in the posterior visual cortex, early in the sensory processing stream, is actually the result of

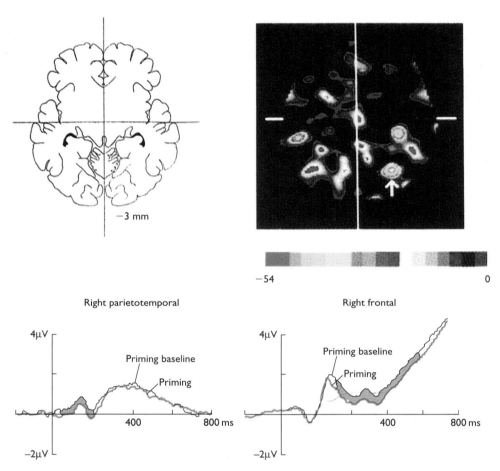

-3 mm

-54   0

Right parietotemporal   Right frontal

4μV   4μV

Priming baseline   Priming baseline

Priming   Priming

400   800 ms   400   800 ms

-2μV   -2μV

Priming in normal subjects using PET imaging and electrical recording. Top: The PET scan at the upper right, corresponding to the horizontal section through the brain at the upper left, shows the average blood flow changes in 15 volunteers. The image was created by subtracting the average image obtained while the volunteers performed a priming task from the average image obtained while they performed a baseline task. The subtraction reveals a reduction in blood flow in right posterior cortex (arrow). Bottom: Average brain waves recorded from the scalps of 40 volunteers while they performed baseline and priming tasks similar to the tasks used in the PET study. The electrical signal related to priming showed a reduction in amplitude in more posterior brain sites within the first 100 milliseconds after stimulus presentation (shaded area on the left). By contrast, changes related to priming could not be detected elsewhere in the brain until after 200 milliseconds.

feedback from higher centers in the brain and that these higher centers are the true locus of the priming effect. To address this issue, Rajendra Badgaiyan and Michael Posner at the University of Oregon carried out a similar study while recording brain electrical activity from 64 scalp electrodes.

The recordings from the scalp electrodes agreed with the PET results. Electrical activity over the posterior cortex was noticeably reduced in the priming condition (when word stems could be completed to form study words), as compared to the baseline condition (when word stems could not form study words). The key finding was that this signature of priming could be detected within 100 milliseconds of the presentation of a word stem. In more forward areas of the brain, no signs of priming were detected until almost 300 milliseconds after word stem presentation. These findings, together with the imaging data, strongly suggest that perceptual priming occurs in the posterior cortex and that information stored in this region about the presented words later allows subjects to process primed word stems more efficiently.

Sensory input apparently makes contact with information in the posterior cortex within 100 milliseconds after a stimulus is presented. Priming may act to reduce the number of responsive neurons and to create a background of relatively silent neurons. The perceptual task might now be handled by a small ensemble of well-tuned neurons, and the result would be a net reduction in neural activity during priming. Thus, it appears that priming occurs within the same cortical visual pathways that are ordinarily involved in perceiving and processing visual information. Neural changes occur within these pathways well before information reaches the memory system of the medial temporal lobe, which is essential for declarative memory. Neural changes that occur at such early stages of processing can be thought of as changes that improve perception itself. Neural

changes that occur after the medial temporal lobe is reached can be thought of as changes that help to create conscious declarative memory.

PERCEPTUAL LEARNING

Whereas perceptual priming occurs after only a single exposure, other kinds of nondeclarative learning within perceptual systems can be more gradual, sometimes unfolding over thousands of practice trials. It was long believed that the early stages of sensory processing were fixed and immutable. According to this view, the areas of cortex that first receive sensory information about the external world serve as "preprocessors," preparing information in a reliable, unvarying way for the more complicated operations that occur at higher levels. Intuitively, this idea seemed eminently reasonable. After all, it is important that we always see a tree as a tree and the face of a friend as the face of that same friend.

This viewpoint has begun to change, in part because studies of perceptual learning suggest that even the earliest stages of sensory processing in the cortex can be changed by experience. "Perceptual learning" refers to an improvement in the ability to discriminate simple perceptual attributes, such as tones or line orientations, simply as the result of performing the discrimination repeatedly. Reward or feedback about errors is not required. In the case of priming, your ability to identify and detect a stimulus improves as the result of having seen the stimulus before. In the case of perceptual learning, you become more expert at discriminating some feature of a stimulus. As we shall see in Chapter 10, training appears to change the structure of the sensory apparatus in the cortex that first receives information from the outside world. Thus, as in the cases of habituation and sensitization studied in *Aplysia*, the ultimate long-term effect of experience is to change the structure of the brain.

The phenomenon of perceptual learning has been studied most extensively in human vision. With practice, people can improve their ability to discriminate texture, direction of motion, line orientation, and many other simple visual attributes. The remarkable feature of this learning is that it is often highly specific to the task and to the specific way in which the training is carried out.

A particularly elegant demonstration of perceptual learning involves the task of discriminating a target from a patterned background. Avi Karni and Dov Sagi at the Weizmann Institute in Israel embedded a small foreground target consisting of three diagonal bars within a large background of horizontal bars. On some trials these three diagonal bars were arranged horizontally, and on other trials the same three diagonal bars were arranged vertically. Volunteers kept their eyes fixed on the center of a computer screen as the display flashed on the screen for a brief 10 milliseconds. After each presentation, subjects decided whether the foreground target was arranged horizontally or vertically. Karni and Sagi controlled the amount of time available to their subjects for visual processing of the display (and its afterimage) by presenting, after a variable delay, a jumbled pattern filled with lines and angles. Accordingly, the available processing time was limited to the interval between the presentation of the display stimulus and the appearance of the jumbled pattern; and the length of this interval controlled the difficulty of the discrimination task. The exact position of the three diagonal bars varied from trial to trial, but they always occupied the same visual quadrant relative to the fixation point (for example, the lower right quadrant in the figure on this page). During each daily session subjects had to perform approximately 1000 decision trials, with rest periods scheduled every 50 trials.

When volunteers began working at this task, they required considerable processing time to make an accurate decision. The volunteer whose

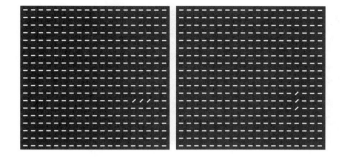

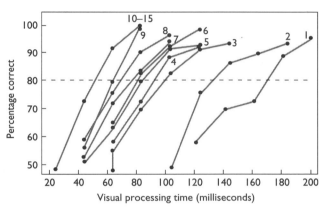

In a test of perceptual learning, subjects saw brief presentations of either the computer-generated display at the upper left or the one at the upper right, then had to decide whether the three diagonal bars in the lower right quadrant of the display were arranged horizontally or vertically. The graph shows the results for one subject who practiced this decision task for 15 days and a total of about 1000 trials per day. The curves, each showing the subject's performance on a different day, demonstrate a remarkable improvement. The leftmost curve is the average performance for days 10 through 15.

performance scores appear in the figure on this page at first required about 180 milliseconds of visual processing time to achieve 90 percent accuracy on the task. When only 50 milliseconds were allowed, he could not do the task at all. However, after 10 training sessions and 10,000 trials of

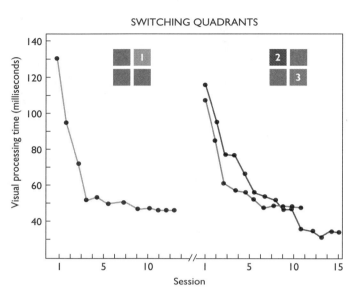

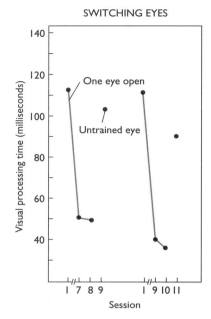

Perceptual learning is very specific. Left: A subject who first learned to discriminate a texture located in the upper-right quadrant (1) of a display could not transfer the skill when the texture was moved to the upper left quadrant (2) and subsequently to the lower right quadrant (3). Each data point shows the amount of processing time needed to perform at 80 percent correct or better. Right: Performance did not transfer between eyes for either of two subjects who were first trained with one eye open and then tested with the untrained eye.

Perceptual learning occurs @ early sensory processing stages

practice spaced across two weeks, his performance was excellent when allowed only 50 milliseconds of processing time. Once he or other subjects had acquired the discrimination ability through practice, they retained it for many weeks.

Further experiments documented the extraordinary specificity of the learning. First, moving the target display to the opposite half-field, or to the untrained quadrant on the trained side, almost completely eliminated the improved discrimination ability, and volunteers had to learn it all over again. Second, when the orientation of the background elements was changed from horizontal to vertical, the discrimination ability did not transfer, and again it had to be relearned. Third, when volunteers kept only one eye open

(monocular training) during the initial training, the learning did not transfer to the other eye.

This remarkable specificity suggests that the learning is occurring at early sensory processing stages in the visual cortex, where the neurons are most sensitive to line orientation and the position of lines within the visual field. At higher levels of visual processing, the neurons process information from both eyes and respond more invariantly across spatial position. If higher visual areas were the locus of perceptual learning, one would have expected the learning to transfer better between eyes and across spatial positions. Accordingly, the most likely locus of perceptual learning is the early visual areas, for example, areas V1 and V2. In these areas, some neurons may perhaps grow longer and branchier axons during perceptual learning, thereby increasing the strength and the number of synaptic connections.

This line of work demonstrates that visual experience has robust and long-lasting effects that, like visual priming, occur within the processing pathways that ordinarily receive visual information. Our visual experiences change the earliest cortical processing stations and affect the way we see. These changes suggest directly how the expert is able to perceive differently from the novice. Thus the landscape painter sees trees differently from the computer programmer, and the portrait painter probably sees faces somewhat differently from the rest of us. A part of this difference is likely the result of genetic makeup, but another important part of it is the result of practice.

All of us, the landscapist and portrait artist included, will make the same basic identifications among the things we see. But the artist will perceive faster, make more finely nuanced comparisons, and see differences more readily. The artist owes these abilities in part to perceptual learning—to changes that have accrued gradually over time in the visual cortex and altered the machinery of perception. Most of these changes are nondeclarative, in the sense that they occur outside of awareness and do not give us conscious remembrances of the past.

EMOTIONAL LEARNING

We can consider priming and perceptual learning mainly as ways that early stages of perceptual processing can become faster, more efficient, and generally more expert as the result of prior experience. But prior experience does not simply improve the speed and efficiency of processing. It can also change the way we feel about what was processed. How we evaluate information—for example, whether we attach positive or negative feeling to a stimulus, our basic likes and dis-

likes—is to a large extent an unconscious (nondeclarative) product of learning. We feel a particular way about a kind of food, a place, or some supposedly neutral stimulus like a tone because of the experiences we have had in association with particular foods, places, and tones.

A compelling demonstration of unconscious learning about likes and dislikes comes from a study of the "mere exposure" effect. Robert Zajonc and his colleagues at the University of Michigan presented pictures of geometric shapes, five times each, to college students at such a fast exposure time (about 1 millisecond for each shape) that the students scarcely had an experience of seeing anything. Indeed, on a later memory test, they could not recognize as familiar the shapes they had seen. Nevertheless, when asked to indicate their preferences, the students preferred the shapes that they had seen over shapes that were new. Thus, subjects developed positive judgments about the material they had seen, even though they had no conscious awareness of having seen the material before. It appears that learning involving emotions can proceed independently of conscious cognition.

The biology of emotional learning has been studied in a number of ways in the laboratory. In one well-studied task, which we encountered in the previous chapter, rats or mice hear a tone and then receive a mild footshock. After only one or two pairings of the tone and the footshock, the animal responds to the tone as if it is afraid. It stops moving, its fur stands up, its blood pressure and heart rate increase—all behaviors that resemble how this animal ordinarily responds to threat or danger. This phenomenon is known as conditioned fear, and it is an example of classical conditioning, the form of learning that we encountered in Chapter 3. Unlike the declarative memories considered in Chapter 5, conditioned fear is not affected by hippocampal lesions. Instead, the learned fear response is eliminated by bilateral damage to the amygdala, a structure in

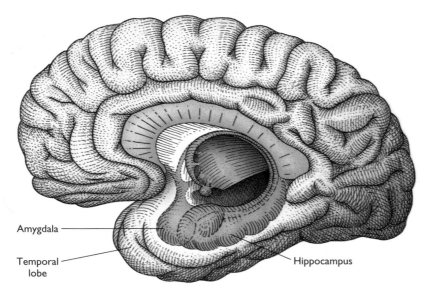

Fear conditioning is associated with Amygdala

Amygdala

Temporal lobe

Hippocampus

Looking at the inside of the human brain from the side, the amygdala is found deep inside the temporal lobe in front of the hippocampus.

the medial temporal lobe immediately in front of the hippocampus.

Studies by Joseph LeDoux at New York University and Michael Davis at Yale have mapped out neural pathways important for fear learning. Exposure to a tone and a shock gives rise to a fear signal. Information about the tone appears to move directly from the sensory areas in the thalamus that first process the tone signal to the basolateral nucleus of the amygdala and to the adjacent perirhinal and insular cortices, which also communicate with the amygdala. The amygdala is composed of more than 10 subregions (or nuclei), and of these the central nucleus is critical for communicating the fear state widely to the many systems that act together to express an organism's response to fear. One system increases heart rate, another causes the freezing of body movements, another slows digestion, and so on.

Interestingly, information travels quickly over a direct pathway from the thalamus to the basolateral nucleus, reaching that structure in about 12 milliseconds. Information about the fear signal also travels a longer route through the cortex, where finer discriminations can be made about

the nature of a stimulus, but information traveling this route takes a little more time to reach the amygdala (19 milliseconds). The more direct pathway provides a way to alert the fear system quickly in the case of danger. Of course, the cortex can modulate the fear response—for example, the response may be dampened if the situation is determined not to be dangerous. The operation of these parallel circuits can explain why we may startle when someone unexpectedly enters the room as we are intently reading a book, even though we realize (almost at the same time) that the person is a friend. Signals reach the amygdala quickly and alert the fear system while the cortex is still fully evaluating them.

Although the neural circuitry of fear learning, including the amygdala and its connections, has been studied most extensively in the rat, the amygdala also plays an important part in fear learning in humans. Antonio Damasio and his colleagues at the University of Iowa made the link to humans by presenting subjects with a neutral tone, followed immediately by a loud (100-dB) noise. After several such pairings, subjects exhibited signs of emotional arousal as soon as the

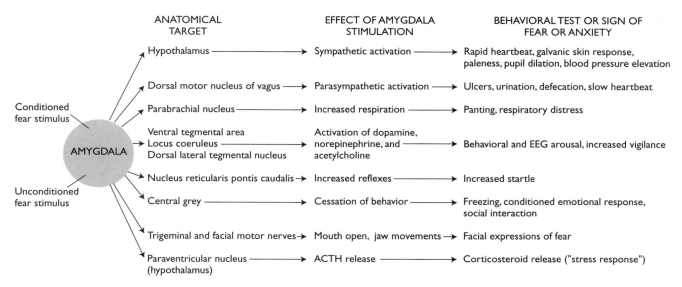

ANATOMICAL TARGET	EFFECT OF AMYGDALA STIMULATION	BEHAVIORAL TEST OR SIGN OF FEAR OR ANXIETY
Hypothalamus	Sympathetic activation	Rapid heartbeat, galvanic skin response, paleness, pupil dilation, blood pressure elevation
Dorsal motor nucleus of vagus	Parasympathetic activation	Ulcers, urination, defecation, slow heartbeat
Parabrachial nucleus	Increased respiration	Panting, respiratory distress
Ventral tegmental area / Locus coeruleus / Dorsal lateral tegmental nucleus	Activation of dopamine, norepinephrine, and acetylcholine	Behavioral and EEG arousal, increased vigilance
Nucleus reticularis pontis caudalis	Increased reflexes	Increased startle
Central grey	Cessation of behavior	Freezing, conditioned emotional response, social interaction
Trigeminal and facial motor nerves	Mouth open, jaw movements	Facial expressions of fear
Paraventricular nucleus (hypothalamus)	ACTH release	Corticosteroid release ("stress response")

Conditioned fear stimulus → AMYGDALA
Unconditioned fear stimulus → AMYGDALA

The central nucleus of the amygdala makes direct connections with a variety of target areas in the brain that express the various symptoms of fear.

neutral tone was presented. Emotional arousal produces small changes in perspiration, which, with small electrodes attached to the fingers, can be recorded as increased skin conductance. Patients with damage to the amygdala did not develop an emotional reaction to the tone. Most of the patients could explain afterward that a tone was regularly followed by a loud noise, but this declarative knowledge about the training situation was not sufficient to make the patients react fearfully or anxiously toward the tone.

Another well-studied task for investigating the neural basis of emotional learning is the "fear-potentiated" startle response. Many species, including humans, startle more vigorously to a loud noise if they are already in a state of fear or arousal, rather than in a calm state, when the noise is presented. We may jump a little (startle) if we hear a loud noise, but we will jump even more if we hear the same noise while we are apprehensively walking alone down a dark alley. Thus, the degree to which we startle can be heightened by fear. To study this phenomenon in

the laboratory rat, an otherwise neutral cue (such as a light) is first paired with a footshock. Then a loud noise is presented either alone or in the presence of the light cue. The startle reflex is greater when the noise is presented together with the light than when the noise occurs alone.

Davis and his colleagues began an analysis of fear-potentiated startle by first working out the neural pathway for the startle reflex itself. When the rat hears a loud noise, changes in the leg muscles can be detected within 6 milliseconds as the animal prepares to jump. The neural pathway that carries information about the noise from the ear to the legs passes through only three synapses in the central nervous system. The nerve fibers from the ear arrive through the auditory nerve at cells in the brain stem (cochlear root neurons; 1 in the figure on the following page) and from there the pathway goes to the pons (reticularis pontis caudalis; 2 in the figure) and then to the spinal cord, where it exits to the muscles. The amygdala is essential if the induction of a fearful state is to heighten the startle response. The

Texture plays an important role in fear conditioning response.

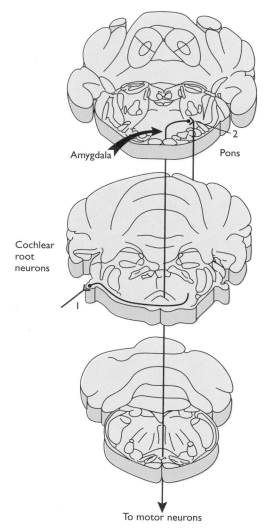

The primary startle circuit in the rat underlies the rapid startle that takes place in response to a loud noise. The circuit consists of cochlear root neurons, which are accessed by the auditory nerve (1); an area in the pons (2); and axons projecting from the pons to motor neurons in the spinal cord. The amygdala enters the startle circuit in the region of the pons.

central nucleus of the amygdala sends out axons that can modulate the primary startle circuit at the level of the pons. Thus, fear is able to modulate the startle response, much as it can modulate heart rate and blood pressure, because the amygdala has a modulating influence on the startle circuit.

Although the amygdala is essential for the development of memories based on fear and other emotions, it is not known whether the memories themselves are actually stored there. The uncertainty stems from the fact that the amygdala is essential not only for acquiring emotional memories, including fear, but also for expressing unlearned emotion. Removal of the amygdala not only abolishes learned fear, it also interferes with the basic capacity to express the emotion of fear. Wild animals become more placid, even fearless, when the amygdala is damaged. Ordinarily, monkeys in captivity retreat from human visitors and hide at the back of their cages. Monkeys with amygdala lesions, however, will approach the front of the cage when visitors are present and even allow themselves to be groomed. Electrical stimulation of the amygdala can also produce a complex pattern of behavior and autonomic changes that resembles fear. In humans, imaging studies have detected changes in activity in the amygdala when people view fearful scenes or when psychiatric patients recollect traumatic events from their past.

One possibility is that the neuronal changes that represent learned fear or some other arousing emotion take place in the neurons of cortical areas, including the perirhinal and insular cortices, that project to the amygdala, and that the amygdala is the structure through which these changes are integrated and broadcast to other brain areas. Another possibility is that the amygdala does in fact store information about positive and negative learned emotional responses. If so, it is likely that only the emotional component of the memory is stored there. Other components, such as the memory of what should be done when the emotional situation repeats itself, are probably stored elsewhere.

The amygdala, then, appears to be at the hub of a neural system that supports nondeclarative memories about emotional events, both positive and negative. This system acquires and perhaps stores the learned emotional response to an object or a situation, a response that is shaped by previous experience. Humans do have preferences and aversions that are inherited through evolution, such as a preference for sweets and an aversion to rapidly approaching, looming objects. However, we also have feelings of like and dislike that result from past experiences: we may be afraid of dogs because a large dog knocked us over as a child; or we may like the sight and sounds of a stream running through deep woods because we vacationed happily in such a place as a child.

These learned feelings of like and dislike work independently of whatever conscious declarative memories we might have about past encounters with particular dogs or vacation spots. To illustrate this point, consider a hypothetical case of a girl of nine who has a bad experience with a stove. Later, as a young adult, two things have happened. First, she may remember the incident quite well. If so, this memory is a declarative recollection, a conscious memory that depends on the hippocampus and related brain structures. Second, independently and in parallel with conscious remembering, this individual may feel a little differently about stoves than the rest of us. She may feel more wary around stoves, or may not like to cook, or may tend to keep a certain distance from stoves. This expression of feeling about stoves reflects the operation of the amygdala. The feeling is a memory to be sure, because it is based on experience, but it is unconscious, nondeclarative, and independent of the capacity for conscious recollection. Because the feeling about stoves and the conscious remembering of what happened are parallel and independent, the existence of this unconscious memory, a fear of stoves, is no guarantee that the young woman can access a declarative memory to explain how the fear came about. The original event may be consciously remembered or it may have been forgotten. Cortical areas are always available to interpret any mental content that is produced, such as a fear or an aversion. But it is then a separate question whether or not the mental experience refers to a memory and, if a memory, whether it is accurate or inaccurate.

Although the amygdala and the hippocampal system independently support nondeclarative emotional memories and declarative memories respectively, the two systems can work together. For example, it is well known that people remember emotionally arousing events especially well. In formal experiments, declarative memory for emotionally arousing material is almost always better than memory for neutral material. This ability of emotion to enhance declarative memory is mediated by the amygdala. James McGaugh and his colleagues at the University of California, Irvine, showed in laboratory animals that mildly stimulating experiences release a variety of hormones into the blood and brain. When these same hormones are injected into animals shortly after they have been trained to perform a simple task, the animals retain the training more strongly. Particularly good examples of this effect come from the injection of stress hormones like epinephrine (adrenalin), ACTH (adrenocorticotropin), and cortisol, which are ordinarily released into the bloodstream when a sudden stress or emergency occurs. These stress hormones appear to exert their effect on memory by activating the amygdala. When the amygdala becomes active, anatomical connections from the amygdala to the cortex may facilitate the processing of whatever stimuli are present. In addition, anatomical connections from the amygdala to the hippocampus could influence declarative memory directly.

Larry Cahill and McGaugh have demonstrated the role of the amygdala in enhancing human memory. Volunteers viewed a slide show while listening to a story narrative. The story, and the slides, told of a boy who is hit by a car and rushed to the hospital for emergency surgery. The volunteers experienced stronger emotional arousal during the central portion of the story (which concerned the accident and the surgery)

Epinephrine, cortisol and ACTH (stress hormones) effect on Amygdala.

and also remembered this part of the story better than the initial and final portions of the story (which concerned relatively neutral events). In addition, memory for the central part of the story was better for these volunteers than for another group of volunteers, who viewed exactly the same slides but heard a different story that interpreted the pictures in a nonemotional way (the boy saw some wrecked cars in a junk yard and witnessed a drill at a hospital).

Patients with lesions restricted to the amygdala remembered the nonemotional parts of the story as well as normal subjects, but they did not have the normal tendency to remember the emotional part of the story better than the other parts. In contrast, Hamann and Squire showed that amnesic patients, who had damage to the hippocampus and related structures but no damage to the amygdala, had poor memory overall for the story but nevertheless exhibited the normal tendency to remember the emotional part of the story best. These results show that the enhancement of memory by emotion results from the amygdala's influence on the declarative memory system.

The importance of the amygdala for emotional remembering was made especially clear in a imaging study of young healthy adults. Cahill, McGaugh, and their colleagues showed either neutral film clips or emotionally distressing film clips to eight volunteers while using positron emission tomography (PET) to measure glucose metabolism in the brain. Brain glucose metabolism is highly correlated with neural activity. Three weeks later and without forewarning, a memory test was given to determine how well the volunteers remembered the film clips. As expected, they recalled the emotionally arousing film clips better than the neutral film clips. The interesting result had to do with metabolic activity in the right amygdala at the time the film clips were first viewed: that activity was highly correlated with the number of emotional film clips that

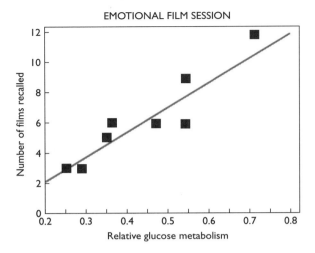

EMOTIONAL FILM SESSION

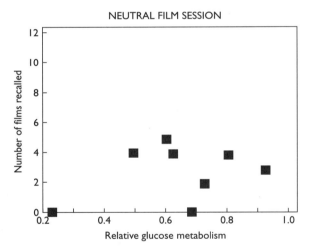

NEUTRAL FILM SESSION

The amygdala and emotional memory. Eight volunteers viewed emotionally arousing or neutral film clips while their brains were imaged by positron emission tomography (PET), which measures glucose metabolic activity, an index of neural activity. Among volunteers who had viewed emotionally arousing film clips, amygdala activity strongly correlated with the ability to recall this material after three weeks (top). Among volunteers who had viewed emotionally neutral film clips, there was no correlation between amygdala activity and the ability to recall the material later (bottom).

volunteers subsequently recalled. Activity in the amygdala did not correlate with the number of neutral films they recalled. Thus, the more active the amygdala at the time of learning, the more it enhanced the storage of those declarative memories that had emotional content.

NONDECLARATIVE MEMORY IN MORAL DEVELOPMENT AND PSYCHOTHERAPY

Studies of early infant-parent interactions have emphasized that infants acquire considerable knowledge about personal relationships and about the world in which they live, and they acquire that knowledge in ways that are nonverbal and unconscious. Recently, the insights from these studies of infants and their caretakers have been applied to the interaction between patients and therapists in psychotherapy. As with infant-parent interactions, it appears that many of the changes that advance the therapeutic process are not attributable to verbal interpretations or to conscious insights but to alterations in unconscious, nondeclarative knowledge. The psychoanalysts Louis Sander, David Stern, and their colleagues at the Process of Change Study Group in Boston have delineated what they call *moments of meeting* in the interactions between patient and therapist—meaningful moments of implicit understanding and trust—which enhance, in an unconscious way, the development of the therapeutic relationship and cause it to move forward to a new level. Thus, contrary to classical psychoanalytic thinking, where interpretation and conscous insights are viewed as being the key events

for progress in the psychotherapeutic process, a *moment of meeting* does not require that unconscious material become conscious. Rather, these moments are thought to lead to lasting changes in behavior by increasing the patient's range of strategies for doing, being, and interacting with others.

The New York psychoanalyst, Marianne Goldberger, has extended this line of thinking to moral development. She has pointed out that people do not generally remember, consciously, the circumstances in which they acquire and assimilate the moral principles that govern their lives. Many of the other dispositions and styles that make up our personality are acquired similarly as nondeclarative knowledge. These principles and dispositions are acquired gradually and almost automatically like the rules of grammar that govern our native language.

Priming, perceptual learning, emotional learning, psychotherapy, and moral development illustrate five ways in which nondeclarative memory can function in parallel with declarative memory. These examples illustrate ways that perception can lead to unconscious memories. The next chapter takes up three other kinds of nondeclarative memory: skill learning, habit learning, and classical conditioning. Some of these involve the motor systems of the brain, which can also record unconscious memories. Finally, some are based not on motor learning but on rather complex perceptual and cognitive abilities. Yet like all forms of nondeclarative memory, they can occur even when we have no awareness about what was learned.

Edvard Munch, Dance on the Shore *(about 1903). Munch (1863–1944) often employed distorted outlines and warm colors to symbolize human emotion or frailty. These dancers on the beach use memory stored as skills and habits to express learned movements and patterns of thought.*

9 | Memory for Skills, Habits, and Conditioning

In 1910, the French philosopher Henri Bergson wrote about what we now call nondeclarative memory. Focusing especially on habits, he wrote, "[It is] a memory profoundly different . . . always bent upon action, seated in the present and looking only to the future. . . . In truth it no longer *represents* our past to us, it *acts* it; and if it still deserves the name of memory, it is not because it conserves bygone images, but because it prolongs their useful effect into the present moment."

When we are introduced to a visitor, we extend our hand in greeting. We can look at the sky on a cloudy day and have some idea about the likelihood of rain. When we read as adults, we execute a complex skill of eye movements and text comprehension that has been improved during thousands of hours of practice. How reliably or accurately we do these things depends on our past experience and on the opportunities we have had for instruction and practice. But we do learn, and we come to perform these and countless other tasks without being aware that we are using memory. In this final chapter on nondeclarative memory, we consider examples of skill learning, habit learning, category learning, and classical conditioning. These extend the discussion of nondeclarative memory in earlier chapters and illustrate its pervasive influence in everyday life.

MOTOR SKILLS

As we noted in Chapter 1, the first hint that memory in the brain is organized into separate systems came from studies of motor skill learning. Patient H.M., though profoundly amnesic for facts and events and ordinary remembering, nevertheless learned to trace the outline of a star in a mirror. Skill learning of this kind has always been easy to understand intuitively as something special, different from the ordinary remembering of recent events. For example, when we learn a new tennis forehand, it seems reasonable to suppose that demonstrating our improved tennis swing is fundamentally different from remembering the tennis lessons themselves or remembering the time we used a forehand to win a particular match.

Our intuitions about motor skills and perceptuomotor skills are quite correct. Learned skills are embedded in procedures, which can be expressed through performance. They are not declarative: one does not need to "declare" anything, and one may not be able, even when pressed, to say very much about what one is doing. Indeed, experience shows that trying to express conscious knowledge about a motor skill while performing it is a good way to impair its execution.

We can learn a motor skill without having any awareness at all of what is being learned.

DBCACBDCBA

A sequence learning task as presented on a computer. Four dashes are always present at the bottom of the screen to denote the four possible locations of an asterisk (A, B, C, or D). During training, the asterisk appears sequentially, moving from one to another of the four locations in a predetermined pattern. Subjects respond to each appearance of the asterisk as rapidly as possible by pressing the key directly beneath the cue.

This curious phenomenon has been best demonstrated in studies of a form of motor skill learning called sequence learning. In the best-studied example of this type of learning, individuals make rapid keypresses in response to a visual signal that appears sequentially in one of four locations on a screen (A, B, C, or D). For example, during 400 consecutive training trials, the visual signal might appear successively in locations DBCACBDCBA until the subjects have been though this sequence 40 times. With practice, subjects gradually learn a sequence of skilled finger movements in response to the changing visual signal. Skill learning is reflected in gradually improved response speed, as subjects begin to predict where the visual signal will appear. However, if the sequence is suddenly changed, performance slows down again. Paul Reber and Squire found that amnesic patients learned this task quite well, though when tested directly the patients exhibited no awareness of the sequence.

In imaging studies, investigators have found that several areas of the brain are specifically acti-

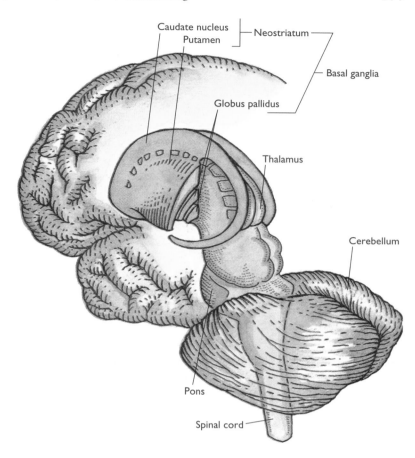

A view of brain structures important for motor skill learning.

vated during sequence learning. These areas include the sensorimotor cortex and two deep structures of the brain known to be involved in motor learning, the caudate nucleus and putamen (which are collectively called the neostriatum). The magnitude of the activation within these regions correlates with how much sequence information a subject has actually acquired. In one study of sequence learning using positron emission tomography (PET), Gregory Berns and his colleagues at the University of Pittsburgh found that activity in the right neostriatum increased when an orderly sequence of numbers was changed during the middle of a test session. Yet participants were unaware of the sequential

structure of the numbers and unaware that the sequence had changed.

Cortical regions are also important during motor skill learning, and these could work together with the neostriatum. Arvi Karni, Leslie Ungerleider, and their colleagues at the National Institute of Mental Health carried out imaging studies of normal volunteers using functional MRI. The volunteers practiced as rapidly as possible a motor skill task in which they touched the thumb with each of the other fingers of the hand in a specified sequence. As would be expected, the area of motor cortex that represents the digits was activated by this task. The interesting finding was that after prolonged training, during which subjects approximately doubled the speed at which they performed the finger-movement sequence, the area of motor cortex that was engaged by the task became larger. This expanded area within the motor cortex then persisted for many weeks, as did the ability to perform the task at increased speed.

It seems likely that practice recruited the activity of additional neurons in the motor cortex, in proportion to the dexterity and speed with which subjects could execute the sequential movements. It is not known where the memory trace of a motor skill is ultimately stored. However, like other examples of nondeclarative memory that we have considered, including the simpler case of *Aplysia,* motor skill learning probably occurs as changes within the circuits already dedicated to performing the skill in question. One possibility is that memory storage occurs within the areas of motor cortex that are engaged during practice. An alternative is that the essential synaptic changes occur in the connections from the cortex to the neostriatum.

An interesting feature of motor skill learning is that there is a shift during learning in which brain systems are important. We have all had the experience of carrying out a well-practiced skill, such as driving a car in a highly automatic way. While driving along a familiar route, we might suddenly notice that we have been proceeding quite satisfactorily for several minutes on automatic pilot as it were, and without paying conscious attention. This experience suggests that areas of the brain involved in attention and awareness may be needed early in skill learning and that these areas become less important as learning proceeds. Steve Petersen and his colleagues at Washington University in St. Louis have documented this phenomenon with imaging studies. Early in learning, the prefrontal cortex tends to be engaged, consistent with its known role in storing information for temporary use. Early learning also engages the parietal cortex, an

area known to be important for visual attention. Other studies show that activity in the prefrontal cortex and parietal cortex also distinguishes subjects who become aware of what they have learned from subjects who do not become aware. Finally, during the earlier stages of motor skill learning, the cerebellum is important. The cerebellum is a large structure lying at the back of the brain, and it is probably necessary for coordinating the specific repertoire of movements that are needed for well-executed, skilled motion and for organizing the timing of these movements. So, it appears that the prefrontal cortex, the parietal cortex, and the cerebellum are all engaged early in motor skill learning. Their combined activity ensures that the correct movements are assembled together and that both attention and working memory are dedicated to the task. After practice with the skill, the prefrontal cortex, the parietal cortex, and the cerebellum all show less activity, and other structures, including the motor cortex and the nearby supplementary motor cortex, become more engaged. These may be structures that together with the neostriatum store the skill-based information in long-term memory and allow the smooth execution of skilled movements.

HABIT LEARNING

To learn a motor skill is to acquire a procedure for operating in the world. The same is true when we learn new habits. As we grow up, we learn to say "please" and "thank you," to wash our hands before meals, and to take on a number of other behaviors, or habits, that are the result of training. We acquire many of these habits early in life, without obvious effort and without taking any special notice that learning is taking place. In this sense, much of habit learning is nondeclarative.

It turns out that the neostriatum is as important for habit learning as it is for the learning of motor skills. In an important study, Mark Packard, Richard Hirsh, and Norman White at McGill University trained rats to perform two different tasks that revealed key differences between habit memory and declarative memory. In one task, animals had to forage for food in the eight arms of a radial maze. On each of several days, animals were placed in the maze and then removed after they had collected a reward from each of the eight maze arms. An error was recorded whenever an animal revisited an arm of the maze in the course of collecting all eight rewards. An efficiently performing rat remembers which arms it has visited and will not repeat entries into the arms it has already visited. Performance on this memory task was disrupted by damage to the *hippocampal* system, but damage to the caudate nucleus had no effect. In a superficially similar task that used the same apparatus, animals had to learn to visit the four arms (out of eight) that were signaled by a light. Only these four lit arms contained a food reward. After about two weeks of training, animals gradually learned to enter the correct arms. In this case, learning was disrupted by damage to the caudate nucleus but not by damage to the hippocampal system. This dissociation between the effects of hippocampal and caudate damage results from the fact that these two tasks, though similar at first glance, are fundamentally different. In performing the foraging task, an animal acquires and uses information about single events; that is, it must remember the specific location it has just visited on a given day. What needs to be remembered is unique to each test day. This type of learning requires the hippocampus. In contrast, the other task is constant from day to day, and the animal must learn about its regularities. Some arms always contain a food reward, and the rat gradually acquires this information through repetition. This *second* task is an example of habit learning.

It has been difficult to study habit learning in human beings because we tend to memorize each

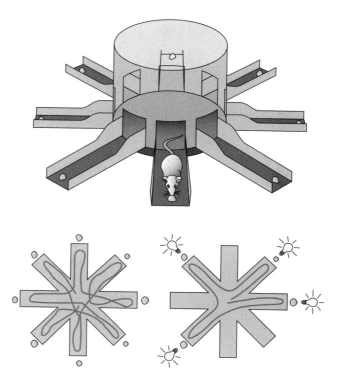

An eight-arm radial maze used to compare declarative memory and habit memory in the rat. In the task of declarative memory, food is available at the end of each arm (lower left). With practice, a normal rat will learn to find all the food by entering each arm just once, following a path like that indicated. In the task of habit memory (lower right), food is available in only four arms, which are illuminated, and the rat learns to visit these arms.

step of a task whenever we can and will engage a declarative memory strategy for tasks that rats or even monkeys learn nondeclaratively as a habit. For example, we could learn which four arms of a maze (out of eight) are the correct ones in just a few trials by quickly memorizing the location of the correct arms. There is no need to build up a habit, because we have available a very efficient declarative memory system, specialized for rapid learning. By contrast, rats can learn only gradually to discriminate the rewarded maze arms from the unrewarded arms. Perhaps rats learn to discriminate maze arms the way humans would learn to discriminate fine wines from mediocre wines or original paintings from forgeries. In these cases, learning proceeds gradually as one learns the relevant dimensons of the problem. Thus, rats learn gradually, without depending on the hippocampus, because they are not memorizing but slowly mastering the problem, as we would a skill.

Scientists can study habit learning in humans if they take special steps to design a task that resists memorization. Barbara Knowlton and Squire used a task developed by Mark Gluck at Rutgers University. The task is presented as a game of predicting the weather. On each trial the subject tries to predict from cards that are presented whether the outcome will be rain or sunshine. Four different cards are used, and on each trial one, two, or three of them can appear. Each card is independently and probabilistically related to the outcome, and the two outcomes occur equally often. For example, one of the four cards is associated with sunshine 75 percent of the time and with rain 25 percent of the time. Another card predicts sunshine 57 percent of the time and rain 43 percent of the time. Each card has its own predictive relationship to the outcome. On each trial, a subject makes a choice of rain or sunshine and immediately receives feedback signaling whether the choice was correct or incorrect. Because of the probabilistic nature of the task, it is impossible to choose correctly all the time. Accordingly, subjects have little success trying to memorize a right answer to each set of cues, and they end up making guesses based on a kind of gut feeling. Most subjects report that they have little sense that they are learning anything at all. Nevertheless, normal subjects do learn about what the cards signify, and they gradually improve their ability to predict the correct outcome of the weather. Amnesic patients learn at the same rate as normal subjects. During 50 training trials, both groups improve from an initial score of 50 percent correct (corresponding to chance

The weather prediction task. Subjects decide on each trial which of two weather outcomes (rain or sunshine) will occur based on a set of one, two, or three cards (out of four possible) that appear on a computer screen.

performance) to about 65 percent correct. Despite their normal performance on the prediction task, the amnesic patients are markedly impaired at answering explicit factual questions about the training episode.

Patients with diseases that affect the caudate nucleus, such as Huntington's disease or Parkinson's disease, are unable to learn this same weather-prediction task. During 50 training trials, these patients chose correctly no better than 53 percent of the time in any 10-trial block. Thus, the caudate nucleus is important for some kinds of habit learning in humans, just as it is in experimental animals.

The importance of the caudate nucleus for the gradual learning of skills and habits can be analyzed at the cellular level. Wolfram Schultz and his colleagues at the University of Fribourg in Switzerland have recorded neuronal responses from awake monkeys in a brain region, the substantia nigra, that provides a major input to the caudate nucleus. Both the caudate and the substantia nigra are affected in Parkinson's disease; both contain a high percentage of neurons that use dopamine as a neurotransmitter. Schultz trained the monkeys to respond to a sound and provided an apple juice reward when the animals responded within two seconds. At the beginning of training, the neurons fired whenever the reward was given. Later, after the monkeys were performing the task well, these same neurons fired in response to the sound, which the monkeys had learned would reliably predict the reward. Thus, dopamine neurons can signal the presence of a reward. These reward signals would be broadcast to higher areas of the brain, like the frontal cortex, that are important for attention and the organization of action. The ability to predict positive feedback is central to the learning process, and dopamine neurons in the neostriatum and related structures are likely to be an important internal mechanism for reward-driven and feedback-driven learning. The caudate and putamen receive overlapping inputs from both sensory cortex and motor cortex, and this double set of inputs could form a basis for associating stimuli and responses. As we saw in the case of classical conditioning in *Aplysia,* one possibility is that habit learning occurs when a modulatory signal (in this case dopamine) produces synaptic changes in cortical neurons that were active just before the reward was presented.

Perceptual and Cognitive Skills

Skill learning is largely concerned with motor skills: how we learn coordinated movements of the hands and feet to meet particular purposes. However, there are also examples of skillful behavior that are not based on learned movements but that nevertheless involve acquiring skillful ways of interacting with the world. When we learn to read our native language, for example, we initially move haltingly from word to word, but after practice we read quickly, moving the eyes to a new location about four times a second and taking the meaning from more than 300 words in one minute. Similarly, when we learn to

program a computer, we first compose and record one command at a time, but eventually we move through mental operations faster than we can type them on the keyboard. These skills are the result of gradual improvement in the perceptual and cognitive procedures that we all use when we perceive, think, and solve problems.

The first example of non-motor skill learning found to be independent of the medial temporal lobe was a reading skill. As we mentioned in Chapter 8, Neal Cohen and Squire showed in 1980 that amnesic patients learned the skill of reading mirror-reversed words at a fully normal rate and retained the skill at a normal level three months later. They were able to learn normally despite the fact that some of the patients did not remember the learning sessions and on formal tests failed to recognize the words they had read.

Skill learning also occurs when we read ordinary text. Gail Musen and Squire exploited the

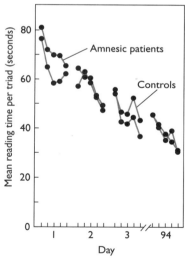

Amnesic patients learned a mirror-reading skill and remembered it at a normal rate when tested three months later. They learned the skill during three daily sessions. On each day, they completed five blocks of trials, and on each trial they saw in succession five unique word triads like that at the top of the graph.

fact that when individuals read aloud repeatedly the same passage of connected prose, the time they require to move through the passage decreases a little with each reading (up to some limit, of course). Thus, with successive readings, the type font, the letters and words, even ideas expressed in the text become easier to perceive. As a result, the material is processed more quickly. However, this improvement in reading speed does not depend on remembering the text in any ordinary sense. Amnesic patients exhibit the same improvement in reading speed as normal subjects, despite the fact that the patients do poorly on memory tests that ask about the content of the passage.

Other examples of skill-based learning have even more of a cognitive flavor than reading skills do. Diane Berry and Donald Broadbent at the University of Cambridge in England used a task that subjects tried to solve on a computer. The subjects imagined that they were running a sugar

production factory and had to decide on each trial how many workers should be hired. The objective was to achieve a particular level of sugar production (9000 tons). The number of workers hired could vary in 12 discrete steps from 100 to 1200, and sugar production could similarly vary from 1000 to 12,000 tons.

At the outset, the computer displayed a starting level of 600 workers and announced that these workers had produced 6000 tons of sugar. Subjects then proceeded to work at the task for 90 trials, deciding on each trial how many workers to hire in order to produce 9000 tons of sugar. The subjects were not told that sugar production was actually determined on each trial by a formula that included the number of workers hired, the previous day's sugar production, and a small random factor.

Squire and Mary Frambach found that amnesic patients improved their performance on this task just as normal subjects did. Both groups

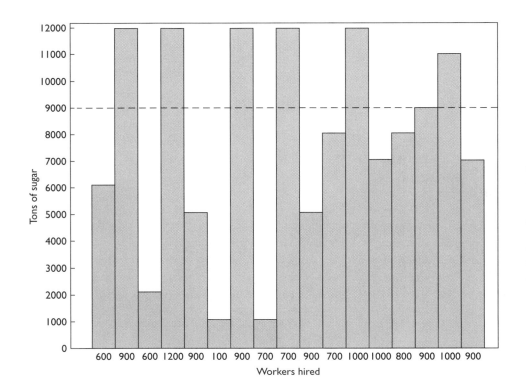

The appearance of the computer screen after a hypothetical series of 12 trials in the sugar-production task. On each trial, the subject decides how many workers to hire with the objective of achieving a production level of 9000 tons of sugar. Production levels of 9000 ± 1000 tons are scored as correct. Thus 3 of the 12 trials shown here would have been scored as correct. In the task itself, subjects can see the number of workers hired only on the most recent trial.

gradually homed in on the correct strategy. All subjects learned not to change the number of workers too abruptly, as this caused sugar production to undershoot or overshoot the target value.

In learning the sugar production task, an individual learns a cognitive skill, which at least in its early stages involves developing a feel for how to do the task. The individual does not actually memorize facts about the task but rather develops a general sense or intuition about how to proceed. This process is nondeclarative. Learning is not accompanied by awareness about how to solve the problem, and learning does not require the brain system that supports declarative memory. Much of what we call "intuition" is probably learned and is based on nondeclarative memory.

LEARNING ABOUT CATEGORIES

A fundamental question about memory concerns how we acquire information about categories and concepts as the result of our encounters with specific instances. When we study a list of items, not only do we learn something about each item in the list, but we also accrue information about what all the items have in common. If we see a display of cups or chairs, for example, we may remember specific cups and individual chairs, but we can also learn about the category of "cupness" and "chairness." Our knowledge of what a cup is, or a chair, is not something taught to us by instruction. Cups and chairs cannot be reduced to an unambiguous set of rules. Rather, our concepts of cup and chair are built up gradually as the result of numerous encounters with different kinds of cups and various kinds of chairs. When one thinks about it, much of our knowledge about the world is in the form of categories. We operate comfortably with categories

every day (birds, cars, buildings, music, and clouds), and appreciating the similarities among the things we see is every bit as important as appreciating the differences.

The question of interest is: What kind of memory underlies the ability to acquire knowledge about categories? The surprising answer is that at least some kinds of category learning are nondeclarative. Category learning can be independent of and parallel to declarative memory, rather than simply derivative from it. People can acquire knowledge about categories implicitly even when their declarative memory for the instances that define the category is impaired.

Knowlton and Squire found that amnesic patients who studied a series of items could extract and retain categorical information about the items as successfully as normal subjects. Knowlton and Squire asked their subjects to study 40 different dot patterns, all variations on the same theme. All the patterns were distortions of a prototype pattern, which subjects did not see, and this prototype pattern was the exact average of all the dot patterns presented for study. Each pattern was presented for five seconds, and in each case the investigators asked the subjects to point to the dot closest to the center of the pattern in order to guarantee that they attended to its global structure.

Five minutes later, subjects were told that these patterns all belonged to a single category of patterns in the same sense that, if a series of different dogs had been presented, they would all belong to the category "dog." Testing then proceeded with 84 new patterns, and for each pattern subjects judged ("yes" or "no") whether it belonged to the same category as the training patterns. The test items consisted of four repetitions of the prototype itself, 40 new distortions of the prototype, and 40 random dot patterns.

Normal subjects were able to categorize these new dot patterns according to how closely they resembled the prototype of the training patterns.

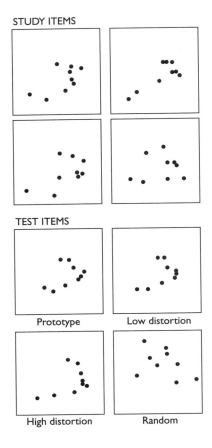

STUDY ITEMS

TEST ITEMS

Prototype

Low distortion

High distortion

Random

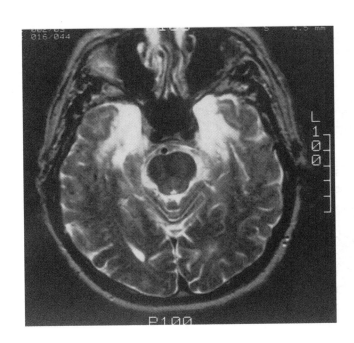

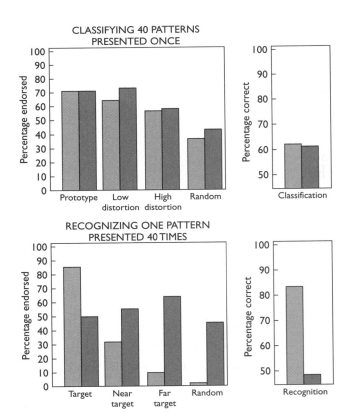

CLASSIFYING 40 PATTERNS PRESENTED ONCE

RECOGNIZING ONE PATTERN PRESENTED 40 TIMES

Learning about categories in amnesia. Top left: Examples of the 40 study items and 84 test items used to assess the learning of dot patterns. The study items were all distortions of a prototype dot pattern that was not presented. The test items were new dot patterns—they could be identical to the prototype, low distortions of the prototype, high distortions of the prototype, or random dot patterns. Top right: A horizontal brain section of amnesic patient E.P. taken with MRI. The area in white showing the extent of damage includes the perirhinal cortex, amygdala, hippocampal formation, and parahippocampal cortex. Bottom: E.P.'s performance is shown by the darker, right-hand bar in each pair of bars. The performance of normal subjects is shown by the left-hand bar in each pair. After studying 40 different training patterns, E.P. and normal subjects endorsed 84 novel dot patterns according to how closely they resembled the prototype of the training patterns. E.P. classified as well as normal subjects. Yet E.P. could not recognize the prototype pattern five minutes after seeing 40 presentations of it. In this case, he was asked to evaluate 84 patterns—4 presentations of the prototype pattern, 20 near targets that resembled the prototype, 20 far targets that resembled it less strongly, and 40 random patterns.

The amnesic patients categorized just as successfully as normal subjects. A particularly striking demonstration of the preserved ability to categorize in amnesia came from an extended study of the severely amnesic patient E.P., whom we described in the opening of Chapter 1. Across six separate tests, E.P. classified about 60 percent of the test items correctly, obtaining virtually the same score as a group of normal volunteers. One might suppose that he succeeded at the categorization task because, despite his poor memory, he could remember a little about the dot patterns and this little bit of memory was enough. After all, he needed only to learn and retain a single concept, the prototype, after seeing 40 related dot patterns.

Further study, however, showed that in fact E.P. is not able to learn even a single concept after 40 study trials if he must use his declarative memory. Specifically, he was unable to retain a single dot pattern in his declarative memory, even after 40 opportunities to become familiar with it. E.P. needed only to pick out the dot pattern that had just been presented 40 times consecutively. Yet, despite the simplicity of this test, he was no better at recognizing the familiar dot pattern than if he had flipped a coin to make his "yes" and "no" choices. He scored 48.4 percent correct, while normal subjects scored 83.3 percent correct.

These results show that aspects of category knowledge can develop in the absence of any detectable memory for the training items. The brain system supporting the ability to categorize must operate in parallel with and independently of the brain system that supports declarative memory. Even when the declarative memory system is inoperative, such that specific items do not persist as individual memories, it is still possible to build up a record of what all the items have in common, to discover and hold onto the regularities that exist among a group of items. Clearly, the medial temporal lobe plays no part in category learning. How does the brain retain knowledge of categories?

One possibility is that when we encounter visual items like dot patterns, the circuitry of the visual cortex gradually changes. These changes in the cortex are a cumulation of the moment-to-moment synaptic changes laid successively on one another. In this way, at any particular moment the pattern of synaptic change provides a running average of cumulative visual experience. This running average would correspond to what all the visual items have in common; that is, it would record the category that the items belong to. If one is to have, in addition to this, a record of the individual dot patterns that have been encountered, then the medial temporal lobe memory system must be intact to work with cortical areas in order to preserve the individual items as separate events.

Reber, Craig Stark, and Squire used functional magnetic resonance imaging to demonstrate the importance of the visual cortex in category learning. They showed 40 related dot patterns to healthy volunteers and then obtained images of brain activity while the volunteers attempted to categorize new patterns. As the volunteers processed categorical patterns, there was a pronounced decrease in activity in the posterior visual cortex. This observation suggests that the subjects processed categorical patterns more easily or quickly, compared to patterns that did not belong to the training category. As we discussed in Chapter 8, studies of priming gave similar results. That is, priming is also associated with a pronounced decrease in activity in posterior cortex. Thus, neural activity appears to decrease either when we see a stimulus a second time, as in priming, or when we see a stimulus that is similar to one presented recently, as in category learning. It seems very likely that category learning of visual stimuli takes place largely in cortical areas that are dedicated to visual information processing. Indeed, it seems likely that perceptual and

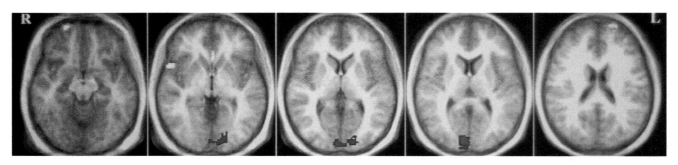

In these horizontal MRI images of the human brain, red and yellow areas indicate where activation increased when subjects processed dot patterns that belonged to the trained category. Blue areas indicate where activation decreased. The main finding was that activity in the posterior visual cortex was reduced during the processing of categorical patterns, in comparison to the activity that occurred during the processing of noncategorical dot patterns. Moving from left to right, the brain sections are from increasingly high levels.

cognitive skill learning, as well as category learning, are cases in which the sensory processing stations themselves change, so as to benefit from the specific perceptual experiences that have occurred in the recent past.

CLASSICAL CONDITIONING OF MOTOR RESPONSES AND NONDECLARATIVE MEMORY

As we described in Chapter 3, classical or Pavlovian conditioning is the most basic and simplest form of associative learning. It occurs when a neutral stimulus precedes a biologically significant stimulus such as food or shock. After many pairings of the two stimuli, the response that is ordinarily made to the second, biologically significant stimulus comes to be elicited reliably by the neutral stimulus. In this way, animals learn about the causal structure of their environment so that their future behavior is better adapted to the circumstances in which they live. The fundamental importance of classical conditioning is indicated by its wide distribution in the animal kingdom. It has been well documented in animals as diverse as invertebrates such as *Drosophila* and *Aplysia,* in lower vertebrates such as fish, in mammals

such as rabbits, rodents, and dogs, and it has been explored extensively in humans as well.

One of the interesting features of classical conditioning is that it can take several forms. The standard procedure for classical conditioning, delay conditioning, is illustrated in the figure on the facing page. In this type of conditioning, the conditioned stimulus is presented first, and it continues to be present while the unconditioned stimulus is presented shortly thereafter. Delay conditioning is relatively reflexive and automatic and is a quintessential example of nondeclarative memory. It is intact both in amnesic patients and in experimental animals with hippocampal lesions. Indeed, work in rabbits showed that the entire forebrain can be removed and delay conditioning still occurs. However, as we shall see, a seemingly minor variation in the training procedure can create a form of classical conditioning, called trace conditioning, that requires the hippocampus, involves awareness, and has other properties of declarative memory.

This story begins with an analysis of a simple form of delay conditioning: the eyeblink response of rabbits. Typically, to condition the eyeblink response, the investigator presents a neutral conditioned stimulus (CS; usually a tone) together with an unconditioned stimulus that causes the rabbit

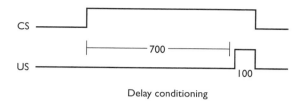

Delay conditioning

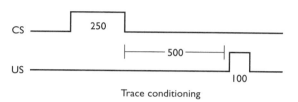

Trace conditioning

Two forms of classical conditioning. The temporal relationship between the conditioned stimulus (CS) and the unconditioned stimulus (US) is different for delayed conditioning and trace conditioning. In a typical delay conditioning procedure, a tone CS is presented and remains on until a 100-millisecond airpuff to the eye (the US) is presented, and both stimuli terminate together. The word "delay" refers to the interval between the onset of the CS and the onset of the US (in this example, about 700 milliseconds). During trace conditioning, the presentation of the CS and the US is separated by an interval (in this example, 500 milliseconds) during which no stimulus is present.

to blink (US; usually an airpuff to the eye). In delay conditioning, the CS is presented just prior to the US. While the CS remains on, the US is presented and the two stimuli terminate together. The temporal relationship between the CS and the US is critical. For defensive conditioning such as the eyeblink response to occur, the CS must precede the US by a fraction of a second. The optimal CS-US interval for the rabbit is 200 to 400 milliseconds. It is similar in *Aplysia* (about 500 milliseconds) and longer in humans. If the interval is longer than about a second, or if the CS occurs at the same time as the US, or if the CS occurs after the US, then conditioning does not take place. Because in classical conditioning two stimuli must be associated, it follows that the conditioning process must occur at brain sites where information about the CS and the US converge. In Chapter 3, we saw that in *Aplysia* the CS and US pathways converge on the sensory neurons of the gill withdrawal circuit.

Work by Richard Thompson and his colleagues at the University of Southern California suggests that the memory trace for classical eyeblink conditioning in the vertebrate is formed and stored in the cerebellum (specifically in the cerebellar cortex and in the interpositus nucleus, a small group of nerve cells lying deep in the cerebellum below the cerebellar cortex). Its unusual anatomy and connectivity help us understand how classical conditioning occurs. The cerebellum receives two major kinds of inputs, both originating in the brain stem at the base of the brain. These are the mossy fibers and the climbing fibers. The mossy fiber input arises mainly from the pons, on the frontward aspect of the brain stem, and from other brain stem sites, and travels to the granule cells of the cerebellum. The axons of the granule cells, which are within the cerebellum, constitute the so-called parallel fibers and they contact the dendrites of the Purkinje cells, also in the cerebellum. The climbing fiber input arises from a cluster of nerve cells in the brain stem called the dorsal accessory olivary nucleus. The axons of these cells synapse directly on the Purkinje cells. Thus, the two inputs to the cerebellum eventually converge on the Purkinje neurons. Each Purkinje cell receives input from many different parallel fibers and from only one climbing fiber. The Purkinje neurons themselves are important because they are the sole source of output from the cerebellar cortex. They exit the cerebellar cortex to synapse on nuclei deep in the cerebellum, including the interpositus nucleus. It is this orderly and well-understood circuitry that has made it possible to prove that the cerebellum is not just important for classical conditioning

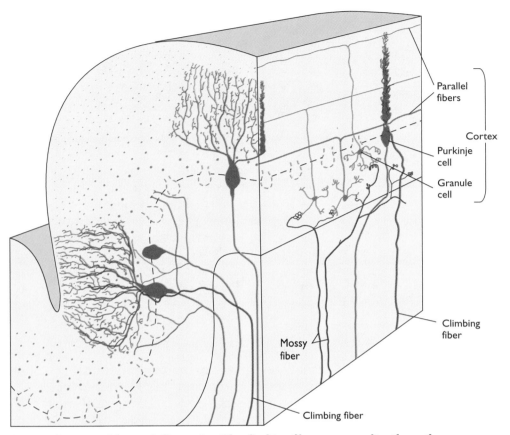

A stereodiagram of the cerebellar cortex. The climbing fibers synapse directly on the Purkinje neurons. The mossy fibers synapse on granule cells (blue) that send out parallel fibers (the blue horizontal lines and the dots at left), which also synapse on Purkinje neurons.

but is also the site where the memory trace is formed and stored.

Working with the rabbit, Thompson and his colleagues first found that small lesions of the interpositus nucleus, involving about 1 cubic millimeter of tissue, completely and permanently prevent a rabbit from learning the eyeblink conditioned response (CR). The same lesion also completely and permanently abolishes retention of a CR that had already been established. In both cases, the unconditioned eyeblink that occurs in response to the airpuff is not affected. This finding shows that the lesion has not affected the ability to perform an eyeblink, only the ability to learn an eyeblink response to a CS.

The importance of the cerebellum is indicated more directly by the fact that electrical stimulation of the two major inputs to the cerebellum can substitute for the tone CS and the airpuff US. Specifically, climbing fibers can be stimulated to evoke an eyeblink, and this electrical stimulation can then serve as a US for eyeblink conditioning. Similarly, stimulation of the mossy fibers can serve as a CS. When climbing fiber and mossy fiber stimulation are paired, mossy fiber stimulation itself comes to evoke an eyeblink. That is, behavioral conditioning occurs.

In subsequent experiments, Thompson, David Lavond, and colleagues explored the role of the cerebellum in storing the memory trace for eyeblink conditioning. They inactivated the interpositus nucleus and the overlying cerebellar cortex by cooling the tissue. Rabbits that had undergone this treatment could not learn the eyeblink CR. After the cooling had worn off, the rabbits learned the CR at the same rate as naive animals. This result shows that the memory trace for the tone-airpuff association is stored in the cerebellum or in downstream structures. The cerebellum is not simply needed to express what has been learned by structures earlier in the flow of information. If that were the case, the CR would have been expressed as soon as the cooling wore off.

In related experiments, the output pathway from the interpositus nucleus (the superior cerebellar peduncle) was inactivated during training by injecting a drug, or one of its main brain stem targets (the red nucleus) was inactivated. In these experiments, a different effect was observed. In this case, the conditioned eyeblink response did not appear during training because motor performance was blocked. However, when the inactivation wore off, the CR was fully evident from the beginning of the test session. This result means that the learning must have occurred upstream from the red nucleus. Memory traces must have been formed during the inactivation, and these memory traces could then be expressed in performance when the inactivation was removed. This body of work provides strong evidence that the

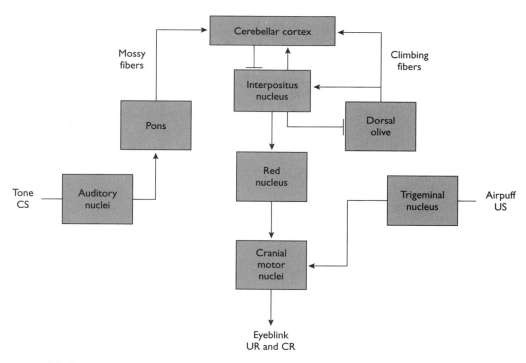

A simplified schematic of the essential brain circuitry involved in eyeblink conditioning. The tone CS enters the circuitry by activating auditory neurons. The airpuff US enters the circuitry by activating neurons in the trigeminal nucleus, which receives tactile information from the skin of the face. The arrows indicate excitatory connections, and the T-junctions indicate inhibitory connections.

essential memory trace for eyeblink conditioning is formed and stored in a small region of the cerebellum. Studies of eyeblink conditioning provide the most complete information currently available about where a memory is located in the vertebrate brain.

Purkinje cells are inhibitory: when they fire, they produce an inhibitory action on cerebellar deep nuclei, including neurons in the interpositus nucleus. Purkinje cells thus act to *decrease* the firing of interpositus neurons. Yet, if conditioning is to increase the frequency of eyeblinks to the conditioned stimulus, one should expect it to result in the *increased* firing of interpositus neurons. This logic implies that Purkinje neurons must decrease their firing during conditioning. In 1982, Masao Ito, then at the University of Tokyo, discovered how this might occur when he discovered the phenomenon of long-term depression (LTD).

LTD is a promising candidate for the synaptic mechanism underlying eyeblink conditioning. LTD occurs when parallel fiber and climbing fiber inputs to the cerebellum are activated in close temporal proximity and at low frequencies (1 to 4 Hz). The result is a decrease in the strength of the parallel fiber synapses onto Purkinje neurons. In simplified preparations, LTD lasts for the duration of the experiment, up to several hours. LTD appears to be mediated entirely postsynaptically, that is, by the Purkinje cell itself. The climbing fiber serves simply to depolarize the Purkinje cell, which allows influx of Ca^{2+}. Roger Tsien and his colleagues at the University of California, San Diego, showed that the parallel fiber stimulation contributes to LTD by generating the gaseous messenger nitric oxide, which then acts to elevate cyclic guanosine monophosphate (cGMP) in the Purkinje cell. cGMP in turn activates a protein kinase (PKG). The result is that Purkinje cells become less responsive to input, presumably due to a reduced sensitivity of their non-NMDA glutamate receptors. When they become less sensitive to input, they decrease their firing and reduce their inhibitory control over interpositus neurons.

The cerebellum is not simply a structure for allowing eyeblink conditioning to occur. Rather, eyeblink conditioning is one example of a discrete motor response, and the classical conditioning of all discrete motor responses is thought to require the cerebellum. In addition, the cerebellum is important for the learning and performance of motor tasks that require the coordination of complex movements. Thus, the cerebellum has a critical role in much of motor learning. Richard Ivry at the University of California, Berkeley, has proposed more broadly that the cerebellum makes a specific contribution to timing, which is important for both motor control and perception. He found that patients with cerebellar lesions failed tasks in which they must judge the temporal interval between pairs of tones. The deficit is not perceptual in itself because the patients had no difficulty judging the relative loudness of two tones. Rather, the cerebellum appears to have a role in timing that includes both the timing of perceptual events and the timing of motor actions. Ito has suggested that the importance of the cerebellum in coordinating motor response may also extend to the coordination of thought itself. In this context, eyeblink conditioning is simply the best-understood example in vertebrates of a learned behavior that requires precise timing, the formation of an association between two events, and the gradual development of coordinated behavior.

CLASSICAL CONDITIONING AND DECLARATIVE MEMORY

The analysis of delay eyeblink conditioning has revealed a very basic and relatively simple form of learning. However, classical conditioning also includes more complex kinds of learning that turn out to have characteristics of declarative

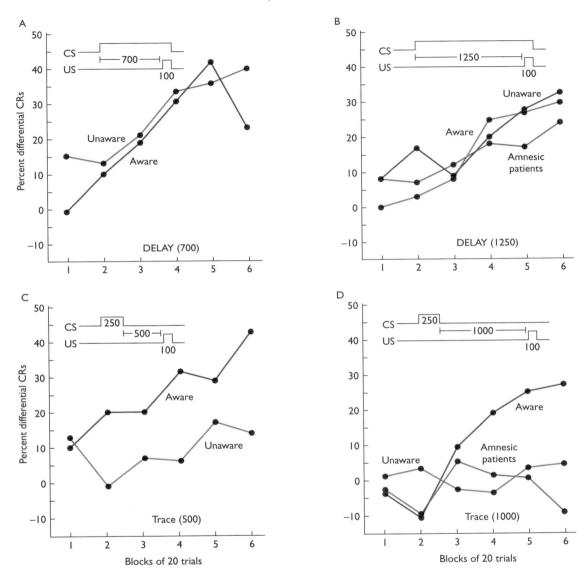

Performance during classical conditioning of the eyeblink response by amnesic patients and four groups of normal volunteers. Each 20-trial block included 10 CS⁺ trials in which a tone (or white noise) occurred together with an airpuff to the eye (the US) and 10 CS⁻ trials in which a tone (or white noise) occurred but the US did not. The data are presented as percent differential conditioned eyeblink responses for each block of 20 trials—that is, percent conditioned responses to the positive conditioned stimulus (CS⁺) minus percent conditioned responses to the negative conditioned stimulus (CS⁻). Only groups that passed a true/false test about the relationship among the CS⁺, the CS⁻, and the US, and were thus considered aware, acquired trace conditioning (bottom graphs). All the groups, whether aware or not, acquired delay conditioning (top graphs).

memory. Consider the case of trace conditioning. This type of conditioning is a variant of classical conditioning in which a short interval (lasting from 500 to 1000 milliseconds) is interposed between the offset of the CS and the onset of the US. Its name comes from the fact that the CS must leave some trace in the nervous system in order for a CS-US association to be established. This small variation creates an entirely new situation, as can be seen from the fact that animals with hippocampal lesions fail to acquire the conditioning. The question is what aspect of trace conditioning requires the hippocampus and why trace conditioning might in fact involve declarative memory.

To address this question, Robert Clark and Squire tested amnesic patients and normal volunteers on two versions of delay conditioning and two versions of trace conditioning, and then assessed the extent to which subjects became aware of the temporal relationships between the CS and the US. Amnesic patients acquired delay conditioning at a normal rate but failed to acquire trace conditioning. The normal volunteers acquired delay conditioning whether or not they were aware of the CS-US relationship, but awareness was a prerequisite for successful trace conditioning. In further experiments, trace conditioning occurred reliably when subjects were told before training about how the CS and US would be related. Conversely, trace conditioning did not occur when subjects could not develop an awareness of the CS-US relationship, because they had to perform an attention-demanding secondary task concurrently with the conditioning trials.

Thus, like other tasks of declarative memory that are impaired after hippocampal lesions, trace conditioning requires individuals to acquire and retain conscious knowledge across a considerable time span (in this case, the 20- to 30-minute conditioning session). A likely reason that trace conditioning requires declarative knowledge is that the trace interval between the CS and the US

makes it difficult to process the CS-US relationship in an automatic, reflexive way. Rather, the situation is probably so complex that the stimuli and their temporal relationship to each other must be represented in the cortex. As discussed in Chapter 5, the hippocampus and related structures then would work jointly with the cortex to establish a usable representation that can persist as memory. People may develop awareness about a task whenever the hippocampus and cortex are engaged during learning. By analogy, those learning and memory tasks that are failed by animals with hippocampal lesions may be tasks for which intact animals must acquire some awareness.

Trace conditioning differs from delay conditioning in its dependence on the hippocampus. However, trace conditioning resembles delay conditioning in that it also depends on the cerebellum. In trace conditioning, just as in delay conditioning, a nondeclarative learning circuit in the cerebellum is required for the generation of a well-timed conditioned response. One possibility is that a representation of the CS-US relationship develops in the cortex, and CS and US information then becomes available to the cerebellum in a format that the cerebellum can use.

Chapters 8 and 9 have illustrated the broad range of nondeclarative forms of learning and memory, and shown how the different forms depend on different brain systems. Priming and perceptual learning are intrinsic to the perceptual machinery of the cortex. Emotional memory requires the amygdala. Skill and habit learning depend crucially on the neostriatum. Classical conditioning of motor responses requires the cerebellum. As we saw in Chapters 2 and 3, many forms of nondeclarative memory are also well developed in invertebrate animals. These forms of learning, such as habituation, sensitization, and classical conditioning, have been preserved through evolutionary history, and they are present in all animals with a sufficiently developed nervous system, from invertebrates such as

Aplysia and *Drosophila* to vertebrates, including humans. Vertebrates, of course, have evolved more complex forms of skill learning and habit learning than invertebrates, corresponding to their more complex perceptual and motor repertoires.

These various forms of nondeclarative memory do not require the participation of the medial temporal lobe memory system. They are ancient in evolutionary terms, they are reliable and consistent, and they provide for myriad unconscious ways of responding to the world. In no small part, by virtue of the unconscious status of these forms of memory, they create much of the mystery of human experience. Here arise the dispositions, habits, and preferences that are inaccessible to conscious recollection but that nevertheless are shaped by past events, influence our behavior and mental life, and are an important part of who we are.

Alberto Giacometti, The Artist's Mother *(1950). Giacometti (1901–1966) painted and sculpted obsessively, creating and destroying each work until he was unable to improve on it. By this means, he endeavored to reproduce from memory an ideal conception of the human form. Here, he depicts his elderly mother.*

10 | Memory and the Biological Basis of Individuality

The reason that we acquire and retain new information so readily is that the systems of the brain that are important for memory are readily modifiable. The synaptic connections within these systems can be strengthened or weakened, and are even capable of permanent structural change. This remarkable plasticity of the brain is fundamental to our individuality and to all aspects of our mental life. As a consequence, the weakening of these capabilities with age or with disease has a profound impact not only on our cognitive functioning but on our very sense of self.

In this final chapter we consider the implications of the brain's capacity for change as it bears on the biological basis of individuality—our sense of self—and on the continuity and maintenance of a free and independent mental life into old age.

THE BIOLOGICAL BASIS OF INDIVIDUALITY

As we acquire new information from day to day and store it as memory, new anatomical changes are thought to be established in the brain. This simple principle has profound implications. Because all of us are brought up in somewhat different environments and have somewhat different experiences, the architecture of each of our brains is modified in *unique* ways. Even identical twins, who share identical genes, will not have identical brains because they are certain to have somewhat different life experiences.

Of course, each of us has the same set of brain structures and a common pattern of synaptic connections that are based on the shared blueprint of our species. This basic blueprint of the human brain—which specifies which region is connected to which, and which class of neurons within each region is connected to which other class of neurons—remains the same within all the individuals of a species. But the details of the blueprint will vary somewhat from person to person. For example, the precise pattern of connec-

tions between neurons and the strength of their connections will differ among individuals according to their genetic makeup. In addition, both the pattern and the strength of synaptic connections will be modified further according to the specific experiences each of us has had.

MODIFYING THE BRAIN THROUGH EXPERIENCE

Dramatic evidence of how pervasively experience can affect the brain comes from studies of perception—how we receive information from the outside world. We experience the outside world through our five senses: touch (and related sensations from the skin), sight, hearing, taste, and smell. Each sensation is first analyzed by appropriate receptors on the body surface and then transmitted through relays to the cortex. Most sensations are thought to come to consciousness in the cerebral cortex.

Modern research on the role of the cerebral cortex in sensation began in 1936 with work on the sense of touch by Phillip Bard, Clinton

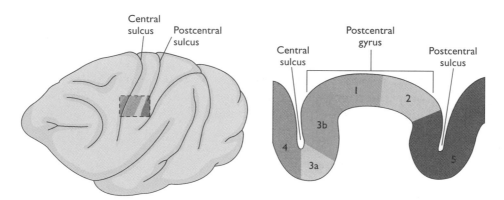

Maps of the body surface are contained in areas 1, 2, 3a, and 3b of the postcentral gyrus, a band of cortex running along the front of the parietal lobe and defined by its position between two sulci, or grooves, the central sulcus and the postcentral sulcus. The square in the monkey brain on the left shows the area of cortex in the postcentral gyrus that is enlarged to the right. Touching a particular area of skin on the body, for example, the hand or foot, elicits neuronal responses in particular locations on the postcentral gyrus.

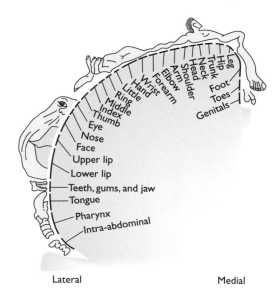

Sensory information from the body surface is received by the postcentral gyrus in the cortex and arranged to form an orderly map. Here the classic map for area 1 in a human being is illustrated. Areas of the body that are important for tactile discrimination—such as the tip of the tongue, the fingers, and the hand—have a disproportionately larger representation, reflecting their more extensive innervation.

Woolsey, and Wade Marshall at Johns Hopkins University. They discovered that the body surface of monkeys was represented on the surface of the brain as a sensory map. Nerves from the skin connect through three synaptic relays with neurons in the postcentral gyrus of the cerebral cortex. The cortical neurons are grouped in an orderly way such that a map of the body is formed. Neighboring areas of the skin end up being represented in neighboring areas of cortex. The neurosurgeon Wilder Penfield soon confirmed the existence of an analogous sensory map in humans, thus establishing the fact that not only monkeys but each of us has within our brain a naked representation of our own body. In our brain, this representation resembles a person—a homunculus—in which the right side of the body is represented in the left hemisphere

and the left side of the body in the right hemisphere.

Until quite recently it had been thought that this representation in the cortex is invariant from one individual to the next and does not change during the life cycle. However, experiments by Michael Merzenich and his colleagues at the University of California, San Francisco, have overturned this idea. They found, surprisingly, that this representation differs significantly from one monkey to the next. This finding raised the question: Are these differences the result of genetic differences between monkeys, or are they the result of learned differences—differences in their tactile experience?

To discover whether tactile experience can alter the cortical representation of the hand area, Merzenich and his colleagues carried out a perceptual learning experiment by teaching monkeys to discriminate between two vibrating stimuli that were applied to a restricted area of the skin on one finger. They trained monkeys for several weeks and for several thousand trials until the animals became quite proficient at the task. They next examined the monkeys to determine how their trained and untrained fingers were represented in cortex. They discovered that the area within the cortex, which represented the stimulated portion of the trained finger, had become more than two times larger than the corresponding area on other fingers. Interestingly, reorganization occurred only in monkeys who actually paid attention to the stimulation. The cortical maps for fingers did not change in monkeys that were passively touched while performing a different task involving an auditory discrimination. It thus appears that, at any moment in the life of an animal, the functional organization of the cortical map for touch is a reflection of the behavioral experiences that the animal has had.

These studies demonstrate that cortical maps representing the body surface are not fixed but dynamic. They are continually modified according to how the sensory pathways are used over

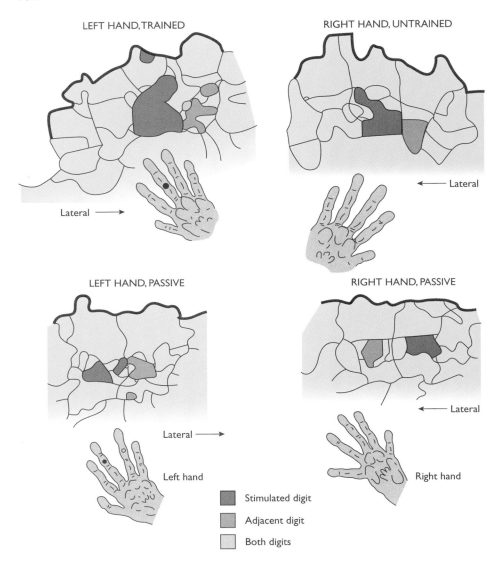

LEFT HAND, TRAINED

RIGHT HAND, UNTRAINED

Lateral →

← Lateral

LEFT HAND, PASSIVE

RIGHT HAND, PASSIVE

Lateral →

← Lateral

Left hand

Right hand

■ Stimulated digit

■ Adjacent digit

□ Both digits

Maps of the hand area from the postcentral gyrus of a monkey. In the drawings of the hands, the black mark shows the area of skin stimulated during behavioral training. Top left: The figure shows the representation in the right postcentral gyrus of the small area of skin trained in the tactile discrimination task as well as the representation of the corresponding location on an adjacent, untrained control finger. Also shown is a small area of cortex that represents both fingers. The top of the map is toward the front of the brain. The arrows point to the side of the brain (lateral). The cortical area representing the trained digit is noticeably larger than the area representing the untrained digit. Top right: Cortical representations in the left postcentral gyrus of the corresponding skin locations for the opposite, unstimulated hand are more nearly equal in size. Bottom: Cortical representations of the skin area stimulated passively in the auditory task (left). Also shown are the representations for the corresponding skin on the opposite, unstimulated hand (right). In these cases, the cortical representation of the stimulated digit did not grow larger.

time. Functional connections can expand and retract as a result of use. Because each of us experiences a different sensory and social environment and because no two people experience exactly the same environment, each of our brains is modified in different ways during our lifetimes. This gradual creation of unique brain architecture provides a biological basis for individuality.

How do these changes come about? In the case of the touch system, the mechanism for change seems to resemble the mechanism used for associative long-term potentiation (LTP) in the hippocampus. (Chapter 6). Areas on the skin that are active at the same time tend to be represented together in the cortical map. Merzenich and his colleagues illustrated this principle by surgically connecting the skin of the third and fourth fingers on one hand. As a result, the connected fingers were always used together, and the inputs from the skin of the two fingers were always correlated, as with associative LTP. This surgical procedure changed dramatically the representation of the hand area in the brain. The normally sharp borders that form the zones representing the third and fourth digits were abolished. Thus, although the zone representing each finger in the cortex is ordinarily neatly demarcated, the pattern and extent of these connections can be modified by experience simply by changing the temporal pattern of input from the fingers.

SKILLS, TALENTS, AND THE DEVELOPING BRAIN

Given the shared evolutionary history of monkeys and humans, it would be surprising if the lessons learned from monkeys did not apply to humans as well. The emergence of technology for brain imaging in the 1990s made it possible to confirm the link directly.

In one study, Thomas Ebert at the University of Konstanz in Germany, and his colleagues, studied the brains of violinists and other string players in comparison to the brains of nonmusicians. String players are an interesting group for studies of how experience affects the brain because during performance the second to fifth fingers of the left hand are manipulated individually and are continuously engaged in skillful behavior. In contrast, the fingers of the right hand, which move the bow, do not express as much patterned, differentiated movement. Brain imaging studies of these musicians revealed that their brains were different from the brains of nonmusicians. Specifically, the cortical representation of the fingers of the left hand, but not of the right, was larger in the musicians. These results dramatically confirm in humans what animal studies had already re-

vealed in more detail. The representation of body parts in the cortex depends on use and on the particular experiences of the individual.

These structural changes are more readily achieved in the early years of life. Thus Wolfgang Amadeus Mozart is Mozart and Michael Jordan is Jordan, not simply because they have the right genes (although genes help) but also because they began practicing the skills they became famous for at a time when their brains were most sensitive to being modified by experience.

Consider the case of Michael Jordan. At the peak of his basketball career Jordan attempted to convert from being one of the best basketball players ever to being a competitive major league baseball player. Jordan had just led the Chicago Bulls to a third consecutive NBA championship. In each of those years, 1990 to 1993, he also led the league in scoring, and in all three years he was voted the most valuable player in the playoffs. However, when now he turned to baseball, at age 31, he could not make the grade despite dedication and serious effort. The Chicago White Sox signed him, but after spring training he was not considered qualified for the major league team or even for their Class AAA farm team. He was sent instead to the Birmingham Barons, the White Sox Class AA team. Here he hit .202 in his only full season, the lowest batting average of any regular player in the league. In so doing he also committed 11 errors in the outfield, which tied him for most in the league.

This outcome should not come as a surprise. Although Michael Jordan had played baseball when he was young, he started playing basketball at the age of eight and soon focused on basketball almost entirely. Playing baseball on a highly competitive level meant learning, honing, and storing new forms of nondeclarative memory, a completely new set of motor and perceptual skills. At age 31, even Michael Jordan could not accomplish that in a few years, and other work suggests that he might never have. Thus, in their study of string players, Ebert and his colleagues

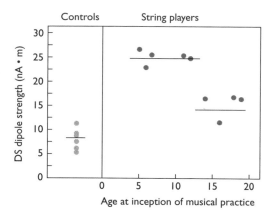

The size of the cortical representation of the fifth finger of the left hand is larger for string players than for nonmusicians. The figure shows the size of cortical representations measured by magnetoencephalography as the dipole strength, which is thought to be an index of total neuronal activity. Among string players, those who begin musical practice before age 13 have a larger representation than do those who begin later.

found that those who learned to play their instruments by the age of 12 years had a larger representation of the fingers of the left hand, their important playing hand, than did those who started later in life. The brain is most modifiable when it is young.

LEARNING AND THE DEVELOPMENTAL FINE TUNING OF BRAIN ARCHITECTURE

As early as the 1960s, Mark Rosenzweig, Edward Bennett, and their colleagues at the University of California, Berkeley, showed that the architecture of the brain can be substantially influenced by the environment in which an animal is raised. Somewhat later, William Greenough at the University of Illinois extended this research. Young rats were placed in large cages with cagemates and toys that were changed frequently. Rats raised in this enriched environment showed an increase in gross cortical weight and thickness.

They also had larger cortical neurons, longer and branchier dendrites, and larger synapses. In the visual cortex, these changes translated into a more than 20 percent increase in the number of synapses per neuron. Thus, animals reared in a complex environment end up with more synapses than do rats reared in the standard laboratory environment.

Is there an indication that these brain changes help the animal in any particular way? The answer is that they do. As an example, rats reared in enriched environments are better at solving complex maze problems. In fact, training rats to learn particular tasks induces structural changes in areas of the brain known to be relevant to the training. Maze training changes the visual cortex, and training on a series of motor coordination tasks changes the cerebellum.

Some of the mechanisms used to change synaptic strength during learning may recapitulate the same mechanisms used to fine tune synaptic connections during development. When the nervous system develops, some of the fine tuning of synaptic connections is thought to be determined by an activity-dependent mechanism similar to long-term potentiation (Chapter 6). It is an intriguing idea that once the developmental program is complete, the capacity to learn depends on an extension of such mechanisms into adulthood, such that the synaptic connections that formed during development can now be strengthened or weakened. Thus, both learning and development may involve activity-dependent changes in the effectiveness of neural connections that result ultimately in anatomical changes in the brain.

At the other end of the life cycle—in old age—memory difficulties are rather common. In some cases, loss of memory is so severe that it leads to a dissolution of the very individuality and sense of identity that developed during a lifetime of learning and memory. If learning and memory are so dependent on brain plasticity, what changes in the brain are related to the various memory difficulties of old age?

MEMORY LOSS AND THE DISSOLUTION OF INDIVIDUALITY

Remind Me Who I Am, Again! This devastating title introduces Linda Grant's book about the progressive memory loss her mother experienced as a result of multi-infarct dementia, a disease that is similar clinically to Alzheimer's disease, even though it has another cause. As her mother's remark implies, memory is the mental glue that binds together and interconnects our life's experiences. Life without the capability to store new information, or to recall previously stored experience, is a life in dissolution, a life without mental past, present, or future, a life without ties to other people or events and, most tragically, without ties to oneself. There is perhaps no more powerful evidence for the importance of memory as the cohesive force in an individual's identity—the sense of self—than the loss of identity that occurs with dementia.

Grant illustrates these points as she writes about a conversation with her mother:

> I tell her that I am going to Poland to track down the family's history, my search for roots.
> 'My parents came from Poland, you know,' she replies.
> 'No, they didn't. That was Dad's family.'
> 'Well, where did mine come from then?'
> 'From Russia, Kiev.'
> 'Did they? I don't remember. Your Auntie Millie will know. Ask her.'
> 'Mum, Auntie Millie died.'
> She begins to cry. 'When? Nobody told me.'
> 'It was years ago, even before Dad died.'
> 'I don't know, I don't remember.'

Grant goes on to comment:

> I don't know if it is a tragedy or a blessing when Jews, who insist on forgiving and forgetting nothing, should end their lives remembering nothing. My mother, the last of her generation, was losing her memory. Only the deep past remained, which emerged at moments, in bits and pieces. . . . This moment, the one she is really living in, is lost from sight as soon as it happens. And the long-ago memories are vanishing too. Only fragments remain. So nearly a century of private history with a cast if not of thousands then of dozens—enough to mount a Broadway musical—is reduced to a shrinking lump of meat weighing a pound or two through which electrical impulses pass. Certain areas of it are permanently turned off at the mains.
> . . . Soon, she will no longer recognize me, her own daughter, and if her disease progresses as Alzheimer's does, her muscles will eventually forget to stay closed against the involuntary release of waste products. She will forget to speak and one day even her heart will lose its memory and forget to beat and she will die. Memory, I have come to understand, is everything, it's life itself.

Alzheimer's disease is the most debilitating and catastrophic memory loss that occurs in the aged, but it fortunately affects only a minority of the aged population. In contrast, a modest weakening of memory function quite distinct from Alzheimer's disease is very common in the elderly. By taking a closer look at, first, age-related memory impairment and, second, Alzheimer's disease, we shall see that both are based on molecular and cellular changes in the brain, providing another link between mind and molecules.

AGE AND DECLINING MEMORY

In reviewing the aging population in the United States, John W. Rowe of Mt. Sinai Medical Center in New York City outlined the remarkable gains in longevity of people in America. In 1900 life expectancy in this country was 47 years and only about 4 percent of the population was over the age of 65. In 1997 life expectancy was more than 76 years and 13 percent of the population was beyond age 65. By the year 2050 life expectancy will likely reach 83 years. People are not only living longer, they are aging better, at least in the United States. Almost 90 percent of those between 65 and 74 claim they have no disability. Even more remarkable, 40 percent of the population over 85 is fully functional. American men who are 65 years old can now expect to spend most of their remaining years living independent lives free of major disability.

The fact that people now live substantially longer than they did 100 years ago has created an important health issue: how to preserve the quality of life for the ever-increasing fraction of the population that is elderly. One of the most pressing problems in normal aging is the gradual appearance of memory problems. As we have seen with athletes and musicians, the ability to modify the connections of the brain varies as a function of age. Older people have, on average, more difficulty achieving these modifications than younger people do. Older people commonly report that their memory is not as good as it used to be. They state, for example, that they may not recall someone's name on a particular occasion, even though that person is well known to them. They may forget where they had placed some familiar item, such as the house keys or a newspaper. The elderly also perform more poorly than younger subjects on a wide variety of memory tests.

Hasker Davis and Kelli Klebe at the University of Colorado, Colorado Springs, gave two memory tests to 469 healthy people aged 20 to 89. In the first test, subjects listened to 15 common words read one at a time and then tried to recall as many words from the list as possible. The same sequence of listening to these words and recalling them was then repeated four more

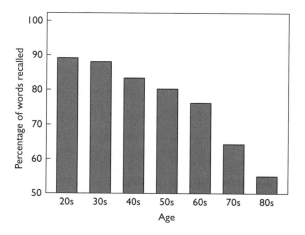

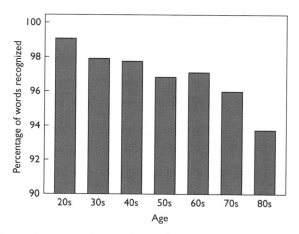

The performance of 469 individuals on two parallel tests of recall and recognition of words. Subjects were tested 20 minutes after learning. The performance of people in their seventies and eighties is dramatically weaker than that of younger people on the recall test. By contrast, there is much less of an age-related difference on the recognition test.

times, with the words always presented in a different order. Finally, 20 minutes after the last recall test, subjects attempted to recall the words once again. The top graph on this page shows that subjects in their seventies and eighties had difficulty recalling words. The same people were also tested on recognition ability using a different list of words and the same learning procedure. Recognition was tested by presenting 30 words, half of which were from the list, and asking people to decide whether each word was from the study list or not. Recognition is much easier than recall, and participants were able to recognize many more words than they could recall. Nevertheless, the elderly scored more poorly than young subjects on both tests.

Normal aging is typically accompanied by a spectrum of cognitive changes, and these include but are not limited to changes in memory. Indeed, there are a number of different abilities that can weaken independently with age. It is for this reason that it is sometimes said that during normal aging people become gradually less alike; they become more distinctive. One common complaint among the elderly is that it is difficult to hear or carry on a conversation when there is competing noise in the background. This phenomenon, one of the earliest-appearing cognitive changes in aging, has been well studied in the laboratory. In one test, subjects hear three pairs of digits in succession, presented through headphones. Each pair is presented to the two ears simultaneously, one number to one ear and the other number to the other ear. Afterward, subjects must report the digits that they heard. Performance on such tasks begins deteriorating as early as age 30 to 40.

The problem is not memory per se, because most people can easily report six digits presented in sequence to one ear. The problem is in processing information presented simultaneously to the two ears. It appears that processing resources are reduced in aging, including the ability to shift rapidly among processing strategies. Weakening of frontal lobe function may be a basis for this problem, as well as for certain other problems encountered by the elderly. These include difficulty remembering where or when a piece of information was acquired (source memory impairment), difficulty remembering the order in which two events occurred (impaired memory for temporal order), and difficulty carrying out intended

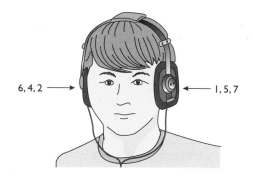

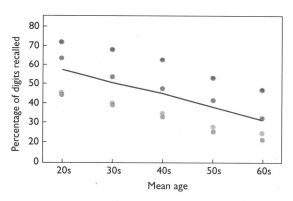

After hearing digits in each ear, a listener tries to recall one set of digits, then the other. Four studies found that performance on this listening task declines steadily with age.

actions at the scheduled time (forgetting to remember).

BENIGN SENESCENT FORGETFULNESS

The difficulty that the elderly have with their memory is often referred to as *benign senescent forgetfulness*, but it is neither completely benign nor does it necessarily begin in senescence. Many of us experience some weakening of memory in the mid-thirties, and troubles with memory typically become more prevalent and pronounced with age. However, memory problems are not universal among the aged. Some aged individuals retain an excellent memory. For example, in the recall test on page 202, nearly 20 percent of people 70 to 79 years old were better at learning and retaining information than the average 30-year-old.

There are, not surprisingly, changes within the brain during the course of aging. Of these, a loss of neurons would provide the most unambiguous indication of reduced function. Unfortunately, the work of counting neurons is notoriously difficult and vulnerable to technical artifacts. The difficulty is that each region of the brain has millions of neurons and it would be impractical to count them all. Instead, the number of neurons is estimated by extrapolating from small samples. In making these extrapolations for any brain region, corrections must be made for the total volume of tissue and for the size of the neurons being counted. Inadequate attention to these factors seems to be the basis for one of the great persisting myths about the human brain: that throughout adulthood we lose an enormous number of neurons—perhaps one hundred thousand—from the brain each day. This widely disseminated idea derives from early mismeasurements of neuronal density in aged brains.

Recently, Mark West at the University of Aarhus in Denmark has applied modern cell-counting techniques to the human brain and found that the loss of neurons with age is in fact quite modest. What about areas of the brain specifically important for declarative memory? West determined that the human hippocampus on each side of the brain contains about 37.3 million neurons. The CA1 region of the hippocampus shows no significant loss with aging, nor does the entorhinal cortex. However, some neurons are lost from parts of the dentate gyrus and from the subiculum. Loss in these areas, which are compo-

nents of the circuitry of the hippocampal formation, could contribute to the forgetfulness that occurs with normal aging. Further, the brain has a number of modulatory systems that are important for attention, learning, and memory. With aging, there is a loss of modulatory input into the hippocampus from brain systems that release the neurotransmitters acetylcholine, norepinephrine, serotonin, and dopamine. Thus, the impairments observed in hippocampal-dependent memory and the concomitant loss of synaptic plasticity may in part arise from loss of modulatory input as well as from structural abnormalities.

It is also clear, however, that one must look beyond the hippocampus to explain all the memory changes in the elderly. Forgetting where you left the car keys or having difficulty learning a new name, a new face, or a new fact could reflect a functional or anatomical change within the hippocampal system. But hippocampal damage cannot explain difficulty in finding words or retrieving the name of an old acquaintance. Amnesic patients, who have hippocampal damage, have no special difficulty with these abilities. These types of impaired memory in the elderly must depend on other changes. For example, changes in the left-hemisphere language areas could explain name-finding problems, and changes in the frontal lobe could explain reduced ability to divide attention between competing sources of information.

ANIMAL MODELS OF AGE-RELATED MEMORY LOSS

Age-related deficits in declarative memory functions have been observed across a wide variety of animal species, including nonhuman primates and rodents. Aged rodents have two characteristics that are reminiscent of elderly humans. First, as in human aging, there is considerable variability in individual performance. In a series of studies carried out by Michela Gallagher at Johns Hopkins University, about 40 percent of aged rats (24 to 27 months old) learned a spatial memory task as quickly and as well as young rats. The remaining 60 percent were slower to learn than the young rats. Second, the effects of aging on learning and memory appeared gradually and were detectable in middle age (14 to 18 months old for the rat).

Through the study of these animals, biologists are coming to understand the aging process in greater detail than could be achieved by studying humans. One important finding is that loss of neurons is not the main change that takes place in the brain with normal aging. For example, studies of mice and rats have documented that there is a loss of synapses in the pathway into the hippocampus, and there are also changes in the capacity of this pathway to generate and maintain long-term potentiation (LTP). Carol Barnes at the University of Arizona first showed that LTP can develop in aged rats to the same magnitude as in young rats, though this level may be reached more gradually. However, the ability to maintain LTP across days is reduced in aging. That is, once induced, LTP decays with abnormal rapidity. This deficit in LTP maintenance is most pronounced in the rats that are the most impaired in their ability to perform behavioral tasks of learning and declarative memory. This work has been extended to the molecular level by Mary Elizabeth Bach and her colleagues at Columbia University, who found that in aged mice there is a defect in the late phase of LTP which is mediated in part by the cAMP-dependent protein kinase. This defect correlates with the severity of memory impairment.

As is the case with rats and humans, aged monkeys exhibit considerable individual variability in their performance on learning and memory tasks. Some perform nearly as well as young monkeys, whereas others resemble young monkeys with surgical damage to the medial temporal lobe or the frontal lobe. In aging, it appears that the changes in the brain that affect the medial temporal lobe and the frontal cortex can occur somewhat independently.

TREATING AGE-RELATED MEMORY IMPAIRMENT

Is there any possibility of treating the frailties of memory that are a common feature of normal aging? It has been known for decades that certain drugs, like amphetamine or caffeine, can enhance cognitive performance, including memory, but they do so mainly by acting as agents that counteract fatigue and boost arousal. As we have come to understand more about the biology of memory, the issue is no longer whether in normal aging some treatment can improve the score on a memory test, but whether memory can be improved beyond what one can do with a good cup of coffee.

Although there has been widespread popular interest in assorted herbs and vitamins for enhancing memory, there are as yet no agreed-upon treatments that will improve memory in healthy individuals. Some positive effects have occasionally been observed in middle-aged and elderly individuals, but they are quite modest.

Recent advances in the biology of memory have further encouraged the search for memory-enhancing drugs. As we saw in Chapter 6, LTP in the hippocampus depends on a novel set of mechanisms (beginning with activation of the NMDA receptor) that are not used for ordinary synaptic transmission. Might it be possible to influence these mechanisms selectively, thereby affecting memory without affecting other neural functions? Of the memory problems associated with aging, the problem of forgetfulness for recently acquired information would seem to be the most ripe for treatment. The ability to retain new memories depends crucially on the hippocampus and related structures, and treatments aimed at these structures, and LTP specifically, would have a rational foundation.

There is an important cautionary note that should be added to any discussion of the search for memory-enhancing drugs. In Chapter 4, we emphasized that forgetting is an important part of remembering. People need to forget or pass over the details in order to grasp the gist, and they need to set aside the details in order to appreciate similarity and metaphor and to form general concepts. Accordingly, the search for memory-enhancing treatments is not a simple matter of wanting to boost retention. Having a maximally powerful memory is not the same as having an optimal, well-functioning memory. However, if a way can be found to combat the forgetfulness that is so common among otherwise healthy elderly individuals without adverse effects, this would certainly be a worthwhile achievement.

THE DEMENTIA OF ALZHEIMER'S DISEASE

Alzheimer's disease, the most common form of dementia, is a neurodegenerative condition that progresses dramatically and inexorably. The disease affects about 10 percent of people between the ages of 65 and 85 and about 40 percent of people over the age of 85. In the United States alone four million people currently suffer from this disease, and over the next 50 years that number is expected to reach 14 million. Thus, Alzheimer's disease is a major public health problem.

The first target of Alzheimer's disease is the entorhinal cortex, the input region to the hippocampus, and the CA1 region of the hippocampus. Both of these regions sustain a significant loss of cells early in the disease. Another prominent site of cell loss is a region at the base of the brain, the *nucleus basalis*, which contains a large population of cholinergic neurons, neurons that use acetylcholine as their chemical transmitter. These cholinergic neurons are modulatory neurons that project widely to the cortex. Loss of these cholinergic neurons can impair attention and other higher mental functions. Although Alzheimer's disease usually begins with memory problems, it progresses to involve intellectual functions very broadly. More of the cortex becomes involved, and patients develop difficulty with language, problem solving, calculation, and judgment. Eventually they lose the very ability to comprehend the world. As the disease worsens and the ability to carry out the activities of daily living becomes increasingly impaired, patients may not be able to move about normally or feed themselves. The course of the disease develops over 5 to 10 years, and because patients become so incapacitated and weakened, they usually die of some other medical illness such as pneumonia.

Since the hippocampal formation is one of the earliest targets of Alzheimer's disease as well as a site of cell loss in ordinary aging, one might suppose that elderly people with age-related memory loss are in fact manifesting early symptoms of Alzheimer's disease. Although some people with age-related memory loss do go on to develop Alzheimer's disease, age-related memory loss is a much more common condition, and the vast majority do not develop dementia. Moreover, as we have seen, the pattern of cell loss in aging is quite different from that observed in Alzheimer's disease. The cell loss characteristic of age-related memory decline occurs in the dentate gyrus and in the subiculum, the output region of the hippocampal system, rather than in the entorhinal cortex and the CA1 region of the hippocampus itself.

Plaques and Tangles: The Hallmarks of Alzheimer's Disease

Alzheimer dementia was first described in 1907 by the German neurologist Alois Alzheimer. He studied a 52-year-old woman who had developed a memory deficit that was accompanied by a progressive loss of cognitive abilities. She died within five years of the onset of her illness. At autopsy Alzheimer discovered the three features that are now recognized as the distinctive diagnostic features of the disease: (1) senile plaques in the brain, particularly in the hippocampus and in the cerebral cortex, (2) neurofibrillary tangles, and (3) loss of neurons.

The *senile plaques* consist of *extracellular* deposits of a protein substance called *amyloid,* surrounded by three cellular elements: (1) dendritic processes of neurons, (2) astrocytes, which are a type of support cell (glial cell) in the brain, and (3) inflammatory cells (microglia). The principal component of the amyloid plaque is a peptide—the amyloid peptide—about 40 amino acids in length. The amyloid peptide is split off from a larger precursor protein called *amyloid precursor protein,* a protein that normally resides in the membrane of nerve cells. Although amyloid precursor protein is normally present in dendrites, cell bodies, and axons of neurons, its function in healthy brains is not yet known. The gene for amyloid precursor protein is located in the midportion of the long arm of human chromosome 21, and a mutation in this gene causes an early-onset form of the disease that is inherited in families.

The *neurofibrillary tangles* are filamentous *intracellular* inclusions that are found in the cell bodies and in the dendrites close to the cell body. These abnormal inclusions are made up of an insoluble form of a normally soluble cellular protein called *tau,* which is part of the cytoskeleton of the cell. The cytoskeleton is essential for maintaining cell geometry and the intracellular trafficking of proteins and organelles, including transport along axons. In affected nerve cells, the cytoskeleton is often abnormal. It is likely that disturbance of the cytoskeleton impairs transport of proteins along the axons to the nerve terminal and thereby compromises the function and viability of neurons. Eventually the affected nerve cells

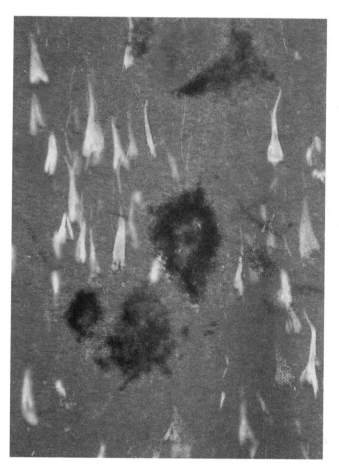

The characteristic amyloid plaques (large dark structures) and neurofibrillary tangles (yellow structures) of Alzheimer's disease are revealed in a histological section of the cerebral cortex taken from a patient with the disease.

PATHWAYS IN THE HIPPOCAMPUS

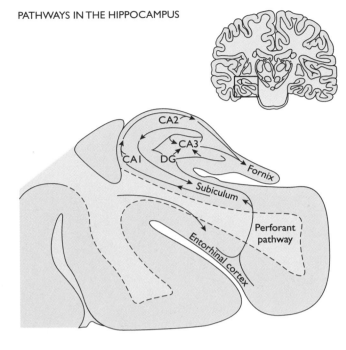

SITES OF NEUROFIBRILLARY TANGLES

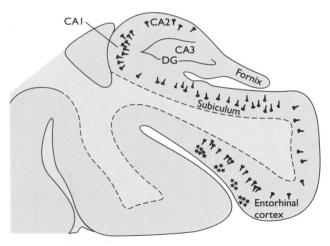

Top: Several major pathways, indicated by arrows, connect the major divisions of the hippocampus. Bottom: Some of these pathways are important sites of neurofibrillary tangle formation in Alzheimer's disease. The tangles are most heavily present in the CA1 region, the entorhinal cortex, and to a lesser degree in the subiculum.

die, and the neurofibrillary tangles are left behind as the tombstones of those cells destroyed by disease. As these neurons die, the synaptic inputs in regions of the brain critical for normal cognitive and memory function are lost.

EARLY- AND LATE-ONSET ALZHEIMER'S DISEASE

Some unfortunate individuals develop Alzheimer's disease in their forties or fifties. These people have the *early-onset form* of the disease, which runs in families and is thought to have a significant genetic component. By studying these families, researchers have identified mutations in several genes that predispose for early-onset Alzheimer's disease. These include mutations in the amyloid precursor protein gene on chromosome 21 (as we mentioned above), in the presenilin 1 gene on chromosome 14, and in the presenilin 2 gene on chromosome 1. All three mutations occur as "autosomal dominant conditions." If one parent has the disease, half the children will also have it.

All three of these mutations lead to abnormal processing of the amyloid precursor protein. As a consequence, with each of these mutations elevated levels of the toxic amyloid peptide are deposited in the brain. These findings have led to the intriguing suggestion that there may only be one major pathogenic mechanism to the genetic forms of Alzheimer's disease. The different mutations cause the same disease by accelerating the rate at which the toxic amyloid peptide is deposited in the brain.

The view that amyloid deposition is an early and critical event in the pathogenesis of Alzheimer's disease is also supported by studies of Down's syndrome. The most common form of mental retardation, Down's syndrome results from the presence of an extra copy of chromosome 21, the very chromosome that carries the gene for the amyloid precursor protein. If people with Down's syndrome live into their thirties, they almost invariably develop Alzheimer's disease, and their brains show the amyloid plaques that are typical of the disease.

The early-onset cases account for only 2 percent of the disease. Nearly 98 percent of patients with Alzheimer's disease exhibit the *late-onset form*. Here the clinical signs of disease first become evident after the age of 60. There are two significant risk factors that predispose for late-onset Alzheimer's disease, both the familial forms and the far more common nonfamilial forms. The first risk factor is the presence of specific alleles of the gene that encodes glycoprotein ApoE, a gene involved in cholesterol storage, transport, and metabolism that is located on the proximal arm of chromosome 19.

As is the case with most genes, the gene for ApoE has several alleles, or variants. ApoE has three alleles called ApoE2, ApoE3, and ApoE4. ApoE3 is the most common allele in the general population. By contrast, ApoE4 is much rarer— about one-fifth as common. However, Allen Roses and his colleagues at Duke University have found that patients with late-onset Alzheimer's disease have ApoE4 at a frequency that is *four* times higher than in the population at large. In fact, if one looks at both copies of chromosome 19 (the one inherited from the father and the one inherited from the mother) and examines the allele of ApoE4 on each copy of that chromosome, one finds that people who carry ApoE4 at both sites are eight times more likely to develop late-onset Alzheimer's disease than is the general population. By contrast, even in families in which some members have late-onset disease, those members who have no copies of ApoE4 have only a small chance—only about one-fifth the chance present in the general population—of developing Alzheimer's disease. These data show

that inheriting a single or a double copy of ApoE4 is a risk factor that increases the likelihood of developing the disease. By contrast, having no copies of ApoE4 is not simply a neutral factor—it actually is a protective factor that reduces the likelihood of developing the disease.

A second, much less well-studied, risk factor is represented by a variant of a gene on chromosome 12, which encodes the protein alpha-2 macroglobulin (A2M). Rudy Tanzi and his colleagues at Harvard University have found a common mutation in this gene that appears to increase greatly the risk for developing the late-onset form of Alzheimer's disease, very much as does the ApoE4 allele. Some estimates suggest that as many as 30 percent of patients with late-onset Alzheimer's disease carry this A2M mutation. Of particular interest is the suggestion that the protein product of the two normal variants of ApoE (ApoE2 and ApoE3), as well as the normal protein product of the A2M gene, are all components of a clean-up or scavenger operation that serves to flush out of the synaptic region various protein fragments, such as the toxic amyloid peptide, that interfere with normal synaptic function. Mutations in the A2M gene, or defects in ApoE, which predispose to late-onset Alzheimer's disease, are thought to do so by interfering with this scavenger function, thereby leading to excessive amyloid deposits in the synaptic region and ultimately to neuronal cell death.

Thus, all five genes so far identified—the mutations in the amyloid precursor protein, the two presenilin mutations, the genetic variant of the ApoE gene (ApoE4), and the mutation of the A2M gene—may all share in common the fact that they participate in a cascade of biochemical steps that serve to either generate or scavenge the amyloid peptide. According to this view, abnormalities in any one of several steps, involved either in production or in scavenging, can lead to

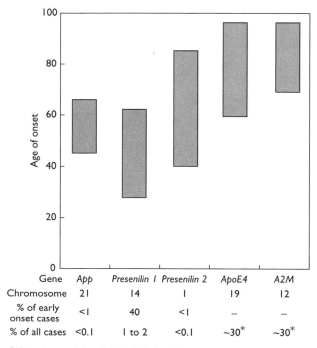

Gene	App	Presenilin 1	Presenilin 2	ApoE4	A2M
Chromosome	21	14	1	19	12
% of early onset cases	<1	40	<1	–	–
% of all cases	<0.1	1 to 2	<0.1	~30*	~30*

* Prevalence of the allele in Alzheimer's cases, where it could potentially be a contributing risk factor.

There are now five genes linked to Alzheimer's disease. Three of these, the amyloid precursor protein (APP) and the two presenilin genes, give rise to mutations that lead to the early-onset form of the disease, which typically can begin in the forties but in occasional cases even earlier. Of these, presenilin 1 is the most common, being mutated in 40 percent of early-onset cases. The late-onset form of Alzheimer's disease is by far the more common. Two genes, the ApoE4 allele and A2M, may each contribute about 30 percent to the total number of cases.

excessive deposition of a toxic amyloid peptide and hence to Alzheimer's disease.

THE SEARCH FOR TREATMENT

Although Alzheimer's disease has no cure, there have been enormous efforts directed toward finding treatments that might slow the progression of the disease. Medical researchers have focused particularly on the cholinergic neurons at the base of the brain that project widely to cortex and that are an early target of the disease. It has not been possible to slow the death of these cells, but drugs are available that can boost the neurotransmitter release of those cells that survive. Two such drugs, tacrine and donepezil, have been shown to have modest therapeutic effects on patients with Alzheimer's disease. For example, in one group of patients tacrine reduced the decline in cognitive function during a six-week treatment period. In another study, which focused on patients with mild or moderately severe disease, tacrine seemed to exert its effect more on attention, and on information-processing speed and accuracy, rather than on memory specifically. Two other drugs, alpha-tocopherol (vitamin E) and selegiline, have been found to delay the progress of the disease by five to eight months. Seligiline boosts the function of some neurotransmitters, while it is thought that vitamin E, because it is an antioxidant, could interrupt the cell-damaging action of amyloid plaques.

Nonsteroidal, anti-inflammatory drugs like aspirin or ibuprofen also exert some protective effect. Middle-aged and elderly persons taking these drugs for two or more years appear to reduce their risk of developing Alzheimer's disease at least for a few years, and perhaps by as much as 25 to 50 percent. Most likely, all these agents only delay the appearance of the disease, not prevent it. Here, too, it is supposed that the treatment might work by protecting neurons from the sequence of degenerative changes that lead to cell death. Particularly promising is the finding that post-menopausal women who take estrogen are less likely to develop Alzheimer's disease than women who do not take estrogen.

As the molecular and cellular basis of Alzheimer's disease becomes better understood, it may be possible to develop rational treatments that target the actual cause of the disease. If so, one can look forward to a time when treatments will not just alleviate symptoms but actually prevent the disease or halt its progression.

The Biology of Memory from Mind to Molecules: A New Synthesis and a New Beginning

In attempting to unravel the workings of memory, we are facing one of the major scientific problems of the twenty-first century and one of its biggest public health challenges. Fortunately, as we have seen, both biology and psychology have undergone great increases in their explanatory power recently. As a consequence, scientists studying memory are now in a much better position to meet this challenge than they were previously.

To begin with, during the last decades, a remarkable unification taking place within the biological sciences has improved our understanding of how genes, cells, and organisms function. For example, because of major advances in our understanding of the gene, we are now able to see how the structure of the gene determines heredity and how the regulation of genes determines development and function. These insights have unified the several, previously independent subfields of biology into a single coherent discipline. Much of what were once separate studies of biochemistry, genetics, cell biology, development, and cancer research have now been united into a coherent field that we call *molecular biology*. The unity achieved by molecular biology has led to an appreciation of the continuity of structure and function that characterizes an organism's cells, a continuity that is apparent as well across all animal species. From this perspective has also come a marvelous sense of the fundamental universality of nature.

Second, an independent and equally profound unification has occurred in the study of mental processes, through the convergence of systems neuroscience and cognitive psychology. This unification has created the coherent discipline that we now call *cognitive neuroscience* and has provided a new perspective on how we perceive, act, learn, and remember.

In this book we have outlined the beginnings of a third unification—a new synthesis—whereby molecular biology and cognitive neuroscience are themselves being combined. This new synthesis—the *molecular biology of cognition*—promises to complete the circle of unification from mind to molecule. Indeed, the study of memory is perhaps the first case of a cognitive process becoming accessible to molecular analysis. The unification of molecular biology and cognitive neuroscience is specifically illuminating the two components of memory that we have considered in this book: the *memory systems* of the brain and the *mechanisms of memory storage*. Compared to what was known even 20 years ago, scientists have learned an extraordinary amount about each of these components.

Consider just three findings central to our current understanding that have emerged from the study of the brain's memory systems. First, memory is not a unitary faculty of the mind but is composed of two fundamental forms: declarative and nondeclarative. Second, each of these two forms has its own logic—conscious recall as compared to unconscious performance. Third, each has its own neural systems.

The molecular study of the mechanisms of memory storage, in turn, has revealed previously unsuspected similarities between declarative and nondeclarative forms of memory. Both forms of memory have a short-term form lasting minutes and a long-term form lasting days or longer. With both forms of memory, the short-term and long-term forms rely on a change in synaptic strength. In both cases, short-term storage calls for only a *transient* change in synaptic strength. Ultimately, in each case, the activation of genes and proteins is necessary for converting short-term to long-term memory. Indeed, both forms of memory storage seem to share a common signaling pathway for activating shared sets of genes and proteins. Finally, both kinds of memory appear to use the growth of new synapses—the growth of both presynaptic terminals and dendritic spines—to stabilize long-term memory.

One further achievement of this new synthesis is the appreciation that we commonly use the memory systems of the brain together. Consider, for example, the viewing of a vase sitting on a table. Our perception of the vase gives rise to a number of different unconscious and conscious effects that can persist as memory. The unconscious memories are particularly diverse. First, the ability to detect and identify the same vase later will be enhanced through the phenomenon of priming. Second, the vase could serve as a cue in the gradual acquisition of a new behavior or habit, shaped by reward. Presentation of the cue (the vase) signals that expressing the behavior will be rewarded. Third, the vase could serve as a conditioned stimulus (CS) and come to elicit a response that is appropriate to coping with an unconditioned stimulus (US) such as a loud noise. Fourth, if the encounter leads to a distinctly pleasant or unpleasant outcome, then one may develop strong positive or negative feelings about the vase. Learning feelings of like or dislike requires the amygdala, learning habits requires the neostriatum, and learning a discrete motor response to a CS requires the cerebellum.

All these memories, potentially triggered by the vase, are unconscious. They are expressed without the experience of any memory content and without the feeling that memory is being used. Moreover, all these memories are the result of cumulative change. Each new moment of experience adds to, or subtracts from, whatever has just preceded. The resulting neural change is the sum of these moment-to-moment changes laid upon each other. There is no sense in which these kinds of memories separate out and store the various individual episodes, each with their own context of time and place, that together make up the cumulative record. In these cases, the vase is a basis for improved perception or a basis for action, but the vase is not remembered as something encountered in the past.

Conscious declarative memory is very different and provides the possibility of re-creating in memory a specific episode from the past. In the case of the vase, we can later recognize it as familiar and also remember the encounter itself, the specific time and place when a unique combination of events converged to create a moment involving this particular vase. For each declarative or nondeclarative memory that might be formed from the encounter with the vase, the starting point is the same distributed set of cortical sites that are engaged when one perceives the vase. Declarative memory uniquely depends on the convergence of input from each of these distributed cortical sites into the medial temporal lobe and ultimately into the hippocampus, and the convergence of this input with other activity that identifies the place and time in which the vase was encountered. This convergence establishes a flexible representation such that the vase can be experienced as familiar and also be remembered as part of a previous episode.

Still another achievement of this synthesis is the finding that, independent of which brain system is recruited in a particular case of learning, the resulting memories are stored as changes in strength at many synapses within a large ensemble of interconnected neurons. Studies of individual synapses in simple nervous systems, like *Aplysia*, have allowed scientists to study close up how individual synapses change as a result of learning. In more complex animals, like the rat and the mouse, we have come to understand in broad outline the molecular events that underlie synaptic change. It appears that, despite the existence of several kinds of memory, synapses may use relatively common mechanisms to accomplish change. Thus, it is not what kinds of molecules are made at synapses that determine what is remembered but rather where and along what pathways the synaptic change occurs. You remember the vase as a familiar vase and your mother's favorite scarf as a familiar scarf not because of the nature of the synaptic change, but because of where those changes are located in the nervous system. The vase and the scarf will have different sites of representation in the brain. Thus, the stored information in its specifics is determined by the *location* of synaptic changes. In contrast, the permanence of that information seems to depend on structural changes that alter the geometry of contact between cells. That is, the architecture of the brain changes to record the effects of experience.

Although much has been learned, all the work to date on the molecular biology and cognitive neuroscience of memory has provided us with just a beginning. We still know relatively little about how and where memory storage occurs. We know in broad outline which brain systems are important for different kinds of memory, but we do not know where the various components of memory storage are actually located and how they interact. We do not, as yet, understand the functions of the various subdivisions of the medial temporal lobe system and how they interact with the rest of the cortex. We also do not understand how declarative information becomes available to conscious awareness. We know almost nothing about how an earlier encounter with a vase is retrieved from memory, or what actually happens as the vase is gradually forgotten, or why it is so easy to confuse a memory with a dream or with something we have only imagined.

Similarly, although we have identified a small number of the genes and proteins that actually switch short-term to long-term memory, we have a long way to go before the molecular steps required for the establishment of the structural changes of long-term memory are completely understood. To date, the most penetrating analyses have been possible only in animals with relatively simple nervous systems, and these animals lack the capacity for declarative memory. Thus, although we now know some things about genes and specific changes important for simple forms of nondeclarative memory, we know much less about complex forms of nondeclarative memory or about declarative memory. LTP holds promise as one mechanism underlying declarative memory, but even this is an artificial phenomenon and we do not know how it is used during normal memory storage.

Some answers to these questions will come from using imaging techniques designed to visualize the human brain while it carries out cognitive tasks of learning, remembering, and forgetting. These experiments will give us correlations between cognitive activities and neural systems for memory. To obtain an understanding of causal mechanisms, scientists can turn to genetic techniques in mice that allow genes to be expressed or eliminated in specific regions and even in specific cells. The most penetrating insights, however, will come from the continued interplay of these two approaches—the molecular biological analysis of cognition and the use of anatomy, physiology, and behavior to analyze the functions of the brain systems that support cognition.

IMPLICATIONS OF A MODERN BIOLOGY OF MEMORY

Because memory is so central to all intellectual activity, the continuing study of memory along the lines we have described can be expected to yield a number of important applications. For example, it is likely that insights into the biology of memory will influence a variety of scholarly disciplines. Some fields, such as philosophy of mind, have already been changed by molecular biological explorations of cognitive processes. Ever since the time of Socrates and the early Platonists, thinkers in each generation have wanted to know: How does experience interact with the mind's innate organization? How do we perceive the world, learn about each other, and remember what we experience? Recently, philosophers such as John Searle at the University of California, Berkeley, and Patricia Churchland at the University of California, San Diego, have used biology to address, in a new way, some of the classical philosophical questions about the nature of mind and conscious experience. They have arrived at novel positions relying not on introspective guesses but on experimental observations from biological studies of cognition.

The study of memory might also affect pedagogy by suggesting new methods of teaching based upon how the brain stores knowledge. For example, recent studies have indicated that different modules of the brain act to process and store different kinds of information. Information about animals and other living things appears to be processed and stored separately from information about inanimate and man-made objects. Can classroom learning be optimized so that it takes advantage of the modular nature of memory storage? Learning in the classroom would optimally proceed so as to minimize interference between adjacent modules and to maximize the ability of modules to work together.

Furthermore, spaced training often gives rise to more effective long-term memory storage than does massed training. This raises several questions: How spaced together in time should one schedule repetitions of the same lesson? Is it better to practice, committing errors as one proceeds? Or is it better to study first and to practice only after the probability of error is low? How many different topics should a student try to study in the same day? One might someday use an evaluation of anatomical changes in the brain as a way of measuring the outcome of novel educational experiments and of remedial programs.

The biology of memory also is likely to have a variety of commercial implications through its impact on technological design. To give but one example, it should have a profound effect on computational science. Earlier studies of artificial intelligence and computer-based pattern recognition indicated that the brain recognizes movement, form, and patterns using strategies that no existing computer begins to approach. To recognize a face as familiar requires a computational achievement that cannot be performed by the same computer that excels at solving logical problems or playing chess. Understanding how the brain recognizes patterns and solves other computational problems related to memory is likely to have a major impact on the design of both computers and robots.

Finally, the modern biology of memory promises to revolutionize medical research and practice in neurology and psychiatry. In neurology, pharmacological treatments for age-related memory loss, and perhaps even Alzheimer's dementia, are becoming a realistic goal. In psychiatry, the biology of memory is likely to have a profound influence on both clinical thinking and therapeutic practice. Insofar as psychotherapy improves mood, attitude, and behavior, the therapy presumably works by producing long-lasting, learning-related structural changes in people's brains. With imaging techniques, it should be possible some day to indicate exactly where and how these changes occur. If this were to happen, it would make the various forms of psychotherapeutic intervention amenable to rigorous scientific scrutiny. In a larger sense, these efforts and these new ways of thought promise to contribute to the intellectual growth of psychiatry as it gradually evolves into an effective medical science that combines its traditional humanistic concerns with new biological insights.

Seen in this light, the emerging synthesis of the molecular biology and cognitive neuroscience of memory that we have described in this book represents both a scientific inquiry of great promise and an aspiration of humanistic and practical scholarship. It is part of the continuous attempt of each generation of scholars and scientists to understand human thought and human action in new and more complex terms. From this perspective, the molecular and cognitive study of memory represents only the most recent attempt, historically, to bridge the sciences, which are traditionally concerned with nature and the physical world, and the humanities, which are traditionally concerned with the nature of human experience, and to use this bridge for the improvement of mentally and neurologically ill patients and for the general betterment of humankind.

Further Readings

Preface

Damasio, A. R. *Descartes' Error: Emotion, Reason, and the Human Brain.* Putnam Publishing Group, New York, 1994.

Descartes, R. (1637). *The Philosophical Works of Descartes,* translated by Elizabeth S. Haldane and G. R. T. Ross, vol. 1. Cambridge University Press, New York, 1970.

Chapter 1

Baddeley, A. *Your Memory: A User's Guide.* MacMillan, New York, 1982.

Kandel, E. R. *The Cellular Basis of Behavior: An Introduction to Behavioral Neurobiology.* W. H. Freeman, San Francisco, 1976.

Schacter, D. L. *Searching for Memory: The Brain, the Mind, and the Past.* Basic Books, New York, 1996.

Scoville, W. B., and B. Milner. Loss of recent memory after bilateral hippocampal lesions. *Journal of Neurology, Neurosurgery and Psychiatry,* 20: 11–21, 1957.

Squire, L. R. (ed.). *Encyclopedia of Learning and Memory.* MacMillan, New York, 1992.

Chapter 2

Bailey, C. H., and M. Chen. Morphological basis of long-term habituation and sensitization in aplysia. *Science,* 220: 91–93, 1983.

Kandel, E. R., J. H. Schwartz, and T. M. Jessell (eds.). *Essentials of Neural Science and Behavior.* Appleton & Lange, Norwalk, Conn., 1995.

Thompson, R. F., and W. A. Spencer. Habituation: A model phenomenon for the study of the neural substrates of behavior. *Psychological Review,* 173: 16–43, 1966.

Tigh, T. J., and R. N. Leighton (eds.). *Habituation: Perspectives from Child Development, Animal Behavior and Neurophysiology.* Erlbaum, Hillsdale, N.J., 1976.

Chapter 3

Davis, R. L. Physiology and biochemistry of Drosophila learning mutants. *Physiological Reviews,* 76: 299–317, 1996.

Kandel, E. R. Small systems of neurons. *Scientific American,* 241: 66–76, 1979.

Chapter 4

Anderson, J. R. *Learning and Memory: An Integrated Approach.* Wiley, New York, 1995.

Druckman, D., and R. A. Bjork (eds.). *In the Mind's Eye: Enhancing Human Performance.* National Academy Press, Washington, 1991.

Druckman, D., and R. A. Bjork (eds.). *Learning, Remembering, Believing: Enhancing Human Performance.* National Academy Press, Washington, 1994.

Neisser, U. *Memory Observed: Remembering in Natural Contexts.* W. H. Freeman, 1982.

Chapter 5

Eichenbaum, H. Declarative memory: Insights from cognitive neurobiology. *Annual Review of Psychology,* 48: 547–572, 1997.

Fuster, J. M. *Memory in the Cerebral Cortex: An Empirical Approach to Neural Networks in the Human and Nonhuman Primate.* The MIT Press, Cambridge, 1995.

Goldman-Rakic, P. S. Working memory and the mind. *Scientific American,* 267: 110–117, 1992.

Milner, B. S., S. Corkin, and H.-L. Teuber. Further analysis of the hippocampal amnesic syndrome: 14 year follow up of H.M. *Neuropsychologia,* 6: 215–234, 1968.

Squire, L. R., and S. Zola-Morgan. The medial temporal lobe memory system. *Science,* 253: 1380–1386, 1991.

Chapter 6

Bliss, T. V., and G. L. Collingridge. A synaptic model of memory: Long-term potentiation in the hippocampus. *Nature,* 361: 31–49, 1993.

Mayford, M., I. M. Mansuy, R. U. Muller, and E. R. Kandel. Memory and behavior: A second generation of genetically modified mice. *Current Biology,* 7: R580–R589, 1997.

Muller, R. A quarter of a century of place cells. *Neuron,* 17: 813–822, 1996.

O'Keefe, J. Place units in the hippocampus of the freely moving rat. *Experimental Neurology,* 51: 78–109, 1976.

Chapter 7

Abel, T., K. C. Martin, D. Bartsch, and E. R. Kandel. Memory suppressor genes: Inhibitory constraints on the storage of long-term memory. *Science,* 279: 338–341, 1998.

Bailey, C. H., D. Bartsch, and E. R. Kandel. Toward a molecular definition of long-term memory storage. *Proceedings of the National Academy of Sciences, USA,* 93: 13445–13452, 1996.

Bailey, C. H., and E. R. Kandel. Structural changes accompanying memory storage. *Annual Review of Physiology,* 55: 397–426, 1993.

Davis, H. P., and L. R. Squire. Protein synthesis and memory: A review. *Psychological Bulletin,* 96: 518–559, 1984.

Tully, T., T. Preat, S. C. Boynton, and M. Del Vecchio. Genetic dissection of consolidated memory in *Drosophila melanogaster. Cell,* 79: 35–47, 1994.

Yin, J. C., M. Del Vecchio, H. Zhou, and T. Tully. CREB as a memory modulator: Induced expression of a dCREB2 activator isoform enhances long-term memory in Drosophila. *Cell,* 81: 107–115, 1995.

Chapter 8

Cahill, L., and J. McGaugh. Mechanisms of emotional arousal and lasting declarative memory. *Trends in Neuroscience,* 21: 294–299, 1998.

Conway, M. *Flashbulb Memories.* Erlbaum, United Kingdom, 1995.

Damasio, A. R. *Descartes' Error: Emotion, Reason, and the Human Brain.* Putnam Publishing Group, New York, 1994.

Davis, M. Neurobiology of fear responses: The role of the amygdala. *Journal of Neuropsychiatry and Clinical Neurosciences,* 9: 382–402, 1997.

LeDoux, J. *The Emotional Brain.* New York: Simon & Schuster, 1996.

The Process of Change Study Group: Stern, D. N., L. W. Sander, J. P. Nahum, A. M. Harrison, K. Lyons-Ruth, A. C. Morgan, N. Bruschweiler-Stern, and E. Z. Tronick. Non-interpretive mechanisms in psychoanalytic therapy: The 'something more' than interpretation. *International Journal of Psycho-Analysis,* 79: 903–921, 1998.

Chapter 9

Grafton, S. T. PET imaging of human motor performance and learning. In: F. Boller and J. Grafman (eds.), *Handbook of Neuropsychology,* Vol. II. Elsevier, New York, 1997.

Petersen, S., H. Van Mier, J. A. Fiez, and M. E. Raichle. The effects of practice on the functional anatomy of task performance. *Proceedings of the National Academy of Science,* 95: 853–860, 1998.

Squire, L. R., and S. M. Zola. Structure and function of declarative and nondeclarative memory systems. *Proceedings of the National Academy of Science,* 93: 13515–13522, 1996.

Chapter 10

Gordon, B. *Memory: Remembering and Forgetting in Everyday Life.* Mastermedia, New York, 1995.

Klawans, H. L. *Why Michael Couldn't Hit and Other Tales of the Neurology of Sports.* W. H. Freeman, New York, 1996.

Marx, J. New gene tied to common form of Alzheimer's. *Science,* 281: 507–509, 1998.

Merzenich, M. M., and K. Sameshine. Cortical plasticity and memory. *Current Opinion in Neurobiology,* 3: 187–196, 1993.

Price, D. L., S. S. Sisodia, and D. R. Borchelt. Alzheimer disease—when and why? *Nature Genetics,* 19: 314–316, 1998.

Rapp, P. R., and D. G. Amaral. Individual differences in the cognitive and neurobiological consequences of normal aging. *Trends in Neuroscience,* 15: 340–345, 1992.

Roses, A. D. Apolipoprotein E and Alzheimers disease. *Scientific American Science Medicine,* 2: 16–25, 1995.

Selkoe, D. J. Aging brain, aging mind. *Scientific American,* 267: 124–142, 1992.

Sources of Illustrations

Facing page 1: Marc Chagall, *Birthday (l'Anniversaire)*, 1915, Oil on cardboard, $31\frac{3}{4}'' \times 39\frac{1}{4}''$ (80.6 × 99.7 cm). The Museum of Modern Art, New York. Acquired through the Lillie B. Bliss Bequest. Photograph ©1999 The Museum of Modern Art, New York. Art ©1999 Artists Rights Society (ARS), New York/ADAGP, Paris.

Page 4: Corbis-Bettmann

Page 6: Department of Experimental Psychology, University of Cambridge, England

Page 8 left: From *The Mill of Thought*, edited by P. Corsi, Electa

Page 8 right: Harvard University Archives

Page 9: From Karl Lashley, *Brain Mechanisms and Intelligence*, Chicago: University of Chicago Press, 1929, Figs. 2 and 28.

Pages 10 and 12: University Relations Office, The McGill Reporter, McGill University

Page 13 top: Courtesy of David Amaral, Ph.D.

Page 13 bottom: Adapted from Brenda Milner, Larry R. Squire, and Eric R. Kandel, Cognitive neuroscience and the study of memory, *Neuron* 20 (1998): 445–468, Fig. 2.

Page 17: Courtesy of Thomas Teyke

Page 18 left: Courtesy of Alfred T. Lamme; from Eric R. Kandel, Small systems of neurons, *Scientific American* 241 (Sept. 1979): 66–76.

Page 18 right: Oliver Meckes/Photo Researchers

Page 19: Institute Archives, California Institute of Technology

Page 20: Photo by James Prince

Page 22: Robert Rauschenberg, *Reservoir*, 1961, National Museum of American Art, Washington, DC, Art Resource, NY. Oil and collage on canvas with objects.

$85\frac{1}{2}'' \times 62\frac{1}{2}'' \times 14\frac{3}{4}''$. ©Robert Rauschenberg/Licensed by VAGA, New York, NY.

Page 25 left: UPI/Corbis-Bettmann

Page 25 right: The Granger Collection, New York

Page 28: Courtesy of Thomas Woolsey, M.D.

Page 32 right: Courtesy of Sir Bernard Katz, Department of Biophysics, University College, London

Page 33: Adapted from T. M. Jessell and E. R. Kandel, Synaptic transmission: A bidirectional and self-modifiable form of cell-cell communication, *Cell* 72/*Neuron* 10 (Jan. 1993 Suppl.): 1–30.

Page 34: Courtesy of Craig Bailey

Page 36: Adapted from R. F. Schmidt, Motor systems. In *Human Physiology*, edited by R. F. Schmidt and G. Thews, translated by M. A. Biederman-Thorson, Springer Berlin, 1983, 81–110.

Pages 37 and 38: Redrawn from Eric R. Kandel, *Cellular Basis of Behavior: An Introduction to Behavioral Neurobiology*, San Francisco: W. H. Freeman, Figs. 9-2 and 7-5.

Page 39: Redrawn from Eric R. Kandel, *A Cell Biological Approach to Learning*, Society of Neuroscience, 1978, p. 21.

Page 41: From V. Castellucci and E. R. Kandel, A quantal analysis of the synaptic depression underlying habituation of the gill-withdrawal reflex in *Aplysia*, *Proceedings of the National Academy of Sciences*, USA 71 (1974): 5004–5008.

Page 43 top (graph): Based on T. J. Carew, H. M. Pinsker, K. Rubinson, and Eric R. Kandel, Physiological and biochemical properties of neuromuscular transmission between identified motoneurons and gill muscle in *Aplysia*, *Journal of Neurophysiology* 37: 1020–1040.

Page 43 middle (recordings and graph):
Adapted from V. Castellucci, T. J. Carew, and E. R. Kandel, Cellular analysis of long-term habituation of the gill-withdrawal reflex of *Aplysia californica*, *Science* 202 (1978): 1306–1308.

Page 43 bottom (graph and diagram):
Adapted from C. H. Bailey and M. Chen, Morphological basis of long-term habituation and sensitization in *Aplysia*, *Science* 220 (1983): 91–93.

Page 46: Jasper Johns, *Zero Through Nine*, 1961, Tate Gallery, London/ Art Resource, NY. Charcoal and pastel on paper. $54\frac{1}{8}''\times 41\frac{5}{8}''$. ©Jasper Johns/ Licensed by VAGA, New York, NY.

Page 50 top: Based on E. R. Kandel, J. H. Schwartz, and T. M. Jessell, *Principles of Neural Science*, 3d ed., New York: Elsevier, 1991, Fig. 65-3A.

Page 50 bottom: Adapted from E. R. Kandel, M. Brunelli, J. Byrne, and V. Castellucci, A common presynaptic locus for the synaptic changes underlying short-term habituation and sensitization of the gill-withdrawal reflex in *Aplysia*, *Cold Spring Harbor Symp. Quant. Biol.* 40 (1976): 465–582.

Page 52: Based on E. R. Kandel, J. H. Schwartz, and T. M. Jessell, *Principles of Neural Science*, 3d ed., New York: Elsevier, 1991, Fig. 12-12.

Page 53 top: After H. Cedar and J. H. Schwartz, Cyclic adenosine monophosphate in the nervous system of *Aplysia californica*: Effect of serotonin and dopamine, *J. Gen Physiology* 60 (1972): 570–587.

Page 53 bottom: Adapted from E. R. Kandel, M. Brunelli, J. Byrne, and V. Castellucci, A common presynaptic locus for the synaptic changes underlying short-term habituation and sensitization of the gill-withdrawal reflex in *Aplysia*, *Cold Spring Harbor Symp. Quant. Biol.* 40 (1976): 465–582.

Page 54: Modified from E. R. Kandel et al., Serotonin, cyclic AMP and the modulation of the calcium current during behavioral arousal. In *Serotonin, Neurotransmission, and Behavior*, edited by A. Gelperin and B. Jacobs, Cambridge, MA: MIT Press, 1991.

Page 55: Based on E. R. Kandel, J. H. Schwartz, and T. M. Jessell, *Principles of Neural Science*, 3d ed., New York: Elsevier, 1991, Fig. 65-3B.

Page 58: The Granger Collection, New York

Page 60: From T. J. Carew et al., Classical conditioning in a simple withdrawal reflex in *Aplysia californica*, *Journal of Neuroscience* 12 (1981): 1426–1437.

Page 61 left: From R. D. Hawkins, T. W. Abrams, T. J. Carew, and E. R. Kandel, A cellular mechanism of classical conditioning in *Aplysia*: Activity-dependent amplification of presynaptic facilitation, *Science* 219 (1983): 400–405.

Pages 61 top right and 63: Based on E. R. Kandel, J. H. Schwartz, and T. M. Jessell, *Principles of Neural Science*, 3d ed., New York: Elsevier, 1991, Figs. 65-7A and 65-8.

Page 65: Based on J. D. Watson et al., *Recombinant DNA*, 2d ed., New York: Scientific American Books, 1992, Fig. 21-12.

Page 68: Milton Avery, *Girl Writing*, 1941, Phillips Collection, Washington, DC. Oil on canvas, $48''\times 32\frac{1}{8}''$. Acquired 1943.

Page 70: Photo by Mary Fox Squire

Page 73: Adapted from Larry R. Squire, *Memory and Brain*, New York: Oxford University Press, 1987, Fig. 32.

Page 77: Adapted from Larry R. Squire, On the course of forgetting in very long-term memory, *Journal of Experimental Psychology: Learning, Memory and Cognition* 15: 241–245 (1989), Fig. 1.

Page 80: Photo by Kevin Walsh

Page 82: Louise Nevelson, *Black Wall*, 1959, The Tate Gallery, London/Art Resource, NY. © 1999 Estate of Louise Nevelson/Artists Rights Society (ARS), New York.

Page 84: The Granger Collection, New York

Page 86: Adapted from Joaquin Fuster, Network memory, *Trends in Neurosciences* 20 (1997): 451–459, Fig. 1.

Page 87: From Leslie G. Ungerleider, Functional brain imaging studies of cortical mechanisms for memory, *Science* 270 (1995): 769, Fig. 1.

Page 89: From Kuniyoshi Sakai and Yasushi Miyashita, Neural organization for the long-term memory of paired associates, *Nature* 354 (1991): 152–155, Figs. 1 and 2.

Page 90: From Alex Martin, Cherli L. Wiggs, Leslie G. Ungerleider, and James V. Haxby, Neural correlates of category-specific knowledge, *Nature* 379 (1996): 649–652, Fig. 1.

Page 93: From Suzanne Corkin, David G. Amaral, R. Gilberto Gonzalez, Keith A. Johnson and Bradley T. Hyman, H.M.'s medial temporal lobe lesion: Findings from magnetic resonance imaging, *Journal of Neuroscience* 17: 3964–3979 (1997), Fig. 1.

Page 95: From Stuart Zola-Morgan, Larry R. Squire, and David G. Amaral, Human amnesia and the medial temporal region: Enduring memory impairment following a bilateral lesion limited to field CA1 of the hippocampus, *Journal of Neuroscience* 6 (1986): 2960–2967, Fig. 7.

Page 97: From Stuart Zola-Morgan and Larry R. Squire, The neuropsychology of memory: Parallel findings in humans and nonhuman primates. In *The Development and Neural Bases of Higher Cognitive Functions, Annals of the New York Academy of Sciences*, 414–456, December 1990, Fig. 3.

Page 98: From R. D. Burwell, W. A. Suzuki, R. Insausti, and D. G. Amaral, Some observations on the perirhinal and parahippocampal cortices in the rat, monkey, and human brains. In *Perception, Memory and Emotion: Frontiers in Neuroscience*, edited by T. Ono, B. L. McNaughton, S. Molotchnikoff, E. T. Rolls, and H. Hishijo, Elsevier UK, 1996, 95–110, Fig. 1.

Page 99: Adapted from Larry R. Squire and Stuart Zola-Morgan, The medial temporal lobe memory system, *Science* 253 (1991): 1380–1386, Fig. 6.

Page 100 top: From Mark F. Bear, Barry W. Connors, and Michael A. Paradiso, *Neuroscience: Exploring the Brain,* Baltimore: Williams & Wilkins, 1996, Fig. 20.23.

Page 100 bottom: Adapted from Michael Bunsey and Howard Eichenbaum, Conservation of hippocampal memory function in rats and humans, *Nature* 379: 255–257 (1996), Fig. 1.

Page 101: Archives of the History of American Psychology, University of Akron, Akron, OH

Page 102: From Larry R. Squire, Pamela C. Slater, and Paul M. Chace, Retrograde amnesia: Temporal gradient in very long term memory following electroconvulsive therapy, *Science* 187: 77–79 (1975), Fig. 1.

Page 103: Adapted from Stuart Zola-Morgan and Larry R. Squire, The primate hippocampal formation: Evidence for a time-limited role in memory storage, *Science* 250: 288–290 (1990), Fig. 2.

Page 104 top: From Pablo Alvarez and Larry R. Squire, Memory consolidation and the medial temporal lobe: A simple network model, *Proceedings of the National Academy of Sciences, USA* 91 (1994): 7041–7045, Fig. 1.

Page 104 bottom: Courtesy of Yasushi Miyashita, Ph.D.

Page 108: Pierre Bonnard, *The Open Window,* 1921. Oil on canvas. $46\frac{1}{2}'' \times 37\frac{3}{4}''$. Acquired 1930, The Phillips Collection, Washington, DC.

Page 112 top: Based on E. R. Kandel, J. H. Schwartz, and T. M. Jessell, *Principles of Neural Science,* 3d ed., New York: Elsevier, 1991, Fig. 65-9.

Page 112 bottom: Adapted from R. A. Nicoll, J. A. Kauer, and R. C. Malenka, The current excitement in long-term potentiation, *Neuron* 1 (1988), 97–103.

Page 114: Adapted from Gustafsson and Wigstrom, 1988, Physiological mechanisms underlying long-term potentiation, *Trends in Neurosciences* 11: 156–162.

Page 115: Modified from V. Y. Bolchakov, H. Odon, E. R. Kandel, and S. Sigelbaum, Recruitment of new sites of synaptic transmission during cAMP-dependent late phase of LTP at CA3-CA1 synapses in the hippocampus, *Cell* 19 (1997): 635–651.

Page 120 left: J. Z. Tsien, P. T. Huerta, and S. Tonegawa, The essential role of hippocampal CA1 NMDA receptor-dependent synaptic plasticity in spatial memory, *Cell* 87 (1996).

Page 121: Based on M. Mayford et al., *Science* 274 (1996): 1678–1683.

Page 122: From J. Z. Tsien et al., The essential role of hippocampal CA1 NMDA receptor-dependent synaptic plasticity in spatial memory, *Cell* 87 (1996).

Page 123: Modified from M. Mayford, I. M. Mansuy, R. U. Muller, and E. R. Kandel, Memory and behavior: A second generation of genetically modified mice, *Current Biology* 7 (1997): R580–R589.

Page 124: Based on R. U. Muller, J. L. Kubie, and J. B. Banack, Jr., Spatial firing patterns of hippocampal complex-spike cells in a fixed environment, *Journal of Neuroscience* 7 (1987): 1935–1950.

Page 125: From Rotenberg et al., Mice expressing activated CaMKII lack low frequency LTP and do not form stable place cells in the CA1 regions of the hippocampus, *Cell* 87 (Dec. 27, 1996): 1351–1361.

Page 128: Roy Lichtenstein, *The Melody Haunts My Reverie*, 1965. ©Estate of Roy Lichtenstein. Licensed by VAGA, New York, NY. Photo by Robert McKeever.

Page 133 and 134: Adapted from W. N. Frost, V. F. Castellucci, R. D. Hawkins, and E. R. Kandel, Monosynaptic connections from the sensory neurons of the gill- and siphon-withdrawal reflex in *Aplysia* participate in the storage of long-term memory for sensitization, *Proceedings of the National Academy of Sciences, USA* 82 (1985): 8266–8269.

Pages 137 and 143: Based on E. R. Kandel, J. H. Schwartz, and T. M. Jessell, *Principles of Neural Science*, 3d ed., New York: Elsevier, 1991, Figs. 12-14 and 36-5.

Page 145: Adapted from C. H. Bailey and M. Chen, Morphological basis of long-term habituation and sensitization in *Aplysia, Science* 220 (1983): 91–93.

Page 147 bottom: From R. A. Nicoll, J. A. Kauer, and R. C. Malenka, The current excitement in long-term potentiation, *Neuron* 1 (1988): 97–103.

Page 148: Modified from V. Y. Bolchakov, H. Odon, E. R. Kandel, and S. Sigelbaum, Recruitment of new sites of synaptic transmission during cAMP-dependent late phase of LTP at CA3-CA1 synapses in the hippocampus, *Cell* 19 (1997): 635–651.

Pages 150, 151, 152, and 153: Ted Abel et al., Genetic demonstration of a role for PKA in the late phase of LTP and in hippocampus-based long-term memory, *Cell* 88 (March 7, 1997): 615–626.

Page 156: Henri Matisse, *Memory of Oceania (Souvenir d'Océanie)*, Nice, summer 1952–early 1953. Gouache and crayon on cut-and-pasted paper over canvas, $9'4'' \times 9'4\frac{7}{8}''$ (284.4 × 286.4 cm). The Museum of Modern Art, New York. Mrs. Simon Guggenheim Fund. Photograph ©1999 The Museum of Modern Art, New York. © 1999 Succession H. Matisse, Paris/Artists Rights Society (ARS), New York.

Page 158: Adapted from Salvatore Aglioti, Joseph F. X. DeSouza, and Melvyn A. Goodale, Size-contrast illusions deceive the eye but not hand, *Current Biology* 5: 679–685 (1995), Fig. 1.

Page 161: Adapted from Stephan B. Hamann and Larry R. Squire, Intact perceptual memory in the absence of conscious memory, *Behavioral Neuroscience* 111: 850–854 (1997), Fig. 2.

Page 163 top: Adapted from Larry R. Squire et al., Activation of the hippocampus in normal humans: A functional anatomical study of memory, *Proceedings of the National Academy of Sciences, USA* 89: 1837–1841 (1992), Fig. 2.

Page 163 bottom: Adapted from Rajendra D. Badgaiyan and Michael I. Posner, Time course of cortical activations in implicit and explicit recall, *Journal of Neuroscience* 17 (1997): 4904–4913, Fig. 2.

Pages 165 and 166: Adapted from Avi Karni and Dov Sagi, Where practice makes perfect in texture discrimination: Evidence for primary visual cortex plasticity, *Proceedings of the National Academy of*

Sciences, USA 88: 4966–4970 (1991), Figs. 1, 2, and 4.

Page 169: From Michael Davis, The role of the amygdala in conditioned fear. In *The Amygdala,* edited by John P. Aggleton, New York: John Wiley & Sons, 1992, 255–305, Fig. 13.

Page 170: Younglim Lee et al., A primary acoustic startle pathway: Obligatory role of cochlear root neurons and the nucleus reticularis pontis caudalis, *Journal of Neuroscience* 16: 3775–3789 (1996), Fig. 12.

Page 172: Larry Cahill et al., Amygdala activity at encoding correlated with long-term, free recall of emotional information, *Proceedings of the National Academy of Sciences, USA* 93: 8016–8021 (1996), Fig. 3.

Page 174: Edward Munch, *Dance on the Beach,* Narodni Galerie, Prague. Erich Lessing/Art Resource, NY. © 1999 Artists Rights Society (ARS), New York.

Page 176: Larry R. Squire and Stuart M. Zola, Structure and function of declarative and nondeclarative memory systems, *Proceedings of the National Academy of Sciences, USA* 93: 13515–13522 (1996), Fig. 3.

Page 179: Adapted from Mark F. Bear, Barry W. Connors, and Michael A. Paradiso, *Neuroscience: Exploring the Brain,* Baltimore: Williams & Wilkins, 1996, Fig. 19.12.

Page 180: From Larry R. Squire and Stuart M. Zola, Structure and function of declarative and nondeclarative memory systems, *Proceedings of the National Academy of Sciences,* USA 93: 13515–13522 (1996), Fig. 2.

Page 181: Adapted from Neal J. Cohen and Larry R. Squire, Preserved learning and retention of pattern-analyzing skill in amnesia: Dissociation of knowing how and knowing that, *Science* 210 (1980): 207–210, Fig. 2.

Page 182: From Larry R. Squire and Mary Frambach, Cognitive skill learning in amnesia, *Psychobiology* 18 (1990): 109–117, Fig. 1.

Page 184: Adapted from Larry R. Squire and Barbara J. Knowlton, Learning about categories in the absence of memory, *Proceedings of the National Academy of Sciences, USA* 92: 12470–12474 (1995), Figs. 2 and 4.

Page 186: From Paul J. Reber, Craig E. L. Stark, and Larry R. Squire, Cortical areas supporting category learning identified using functional MRI, *Proceedings of the National Academy of Sciences, USA* 95: 747–750 (1998), Fig. 3.

Page 187: Adapted from Robert E. Clark and Larry R. Squire, Classical conditioning and brain systems: The role of awareness, *Science* 280: 77–81 (1998), Fig. 1.

Page 188: Adapted from John C. Eccles, Masao Ito, and Janos Szentagothai, *The Cerebellum as a Neuronal Machine,* New York: Springer-Verlag, 1967, Fig. 1.

Page 189: Adapted from Richard F. Thompson and David J. Krupa, Organization of memory traces in the mammalian brain, *Annual Review of Neuroscience* 17: 519–550 (1994), Fig. 1.

Page 191: From Robert E. Clark and Larry R. Squire, Classical conditioning and brain systems: The role of awareness, *Science* 280 (1998): 77–81, Fig. 3.

Page 194: Alberto Giacometti, *The Artist's Mother,* 1950. Oil on canvas, $35\frac{3}{4}'' \times 24''$ (89.9×61 cm). The Museum of Modern Art. Acquired through the Lillie P. Bliss Bequest. Photograph ©1999 The Museum of Modern Art, New York. © 1999 Artists Rights Society (ARS), New York/ADAGP, Paris.

Page 196: Adapted from E. R. Kandel, J. H. Schwartz, and T. M. Jessell, *Principles of Neural Science,* New York: Elsevier, 1991, Fig. 26-4.

Page 197: From Wilder Penfield and Theodore Rasmussen, *The Cerebral Cortex of Man: A Clinical Study of Localization of Function,* New York: Macmillan, 1952, Fig. 17.

Page 198: Adapted from Gregg H. Recanzone et al., Topographic reorganization of the hand representation in cortical area 3b of owl monkeys trained in a frequency-discrimination task, *Journal of Neurophysiology* 67: 1492 (1992), Figs. 8, 9, and 10.

Page 200: From Thomas Elbert et al., Increased cortical representation of the fingers of the left hand in string players, *Science* 270 (1995): 305–307, Fig. 1(b).

Page 202: Courtesy of Hasker Davis and Kelli Klebe

Page 203: Adapted from Robin A. Barr, Some remarks on the time-course of aging. In *New Directions in Memory and Aging,* edited by Leonard W. Poon, James L. Fozard, Laird S. Cermak, David Arenberg, and Larry W. Thompson, Hillsdale, NJ: Lawrence Erlbaum Associates, 1980, Fig. 9.1.

Page 207: Courtesy of Gary W. Van Hoesen, Ana Solodkin, and Paul Reimann of the Departments of Anatomy and Neurology at the University of Iowa College of Medicine.

Page 208: Based on E. R. Kandel, J. H. Schwartz, and T. M. Jessell, *Principles of Neural Science,* 3d ed., New York: Elsevier, 1991, Fig. 62-4.

Page 210: Adapted from Jean Marx, *Science* 281 (1998): 507–509.

Index

Selected Books in the Scientific American Library Series

If you would like to purchase additional volumes in the
Scientific American Library, please send your order to:

Scientific American Library
41 Madison Avenue
New York, NY   10010